W0112058

The History and Evolution of the North
American Wildlife Conservation Model

Robert E. Wright

The History and Evolution of the North American Wildlife Conservation Model

palgrave
macmillan

Robert E. Wright
American Institute for Economic Research
Massachusetts, MA, USA

ISBN 978-3-031-06162-2   ISBN 978-3-031-06163-9   (eBook)
https://doi.org/10.1007/978-3-031-06163-9

© The Author(s), under exclusive licence to Springer Nature Switzerland AG 2022
This work is subject to copyright. All rights are solely and exclusively licensed by the Publisher, whether the whole or part of the material is concerned, specifically the rights of translation, reprinting, reuse of illustrations, recitation, broadcasting, reproduction on microfilms or in any other physical way, and transmission or information storage and retrieval, electronic adaptation, computer software, or by similar or dissimilar methodology now known or hereafter developed.
The use of general descriptive names, registered names, trademarks, service marks, etc. in this publication does not imply, even in the absence of a specific statement, that such names are exempt from the relevant protective laws and regulations and therefore free for general use.
The publisher, the authors and the editors are safe to assume that the advice and information in this book are believed to be true and accurate at the date of publication. Neither the publisher nor the authors or the editors give a warranty, expressed or implied, with respect to the material contained herein or for any errors or omissions that may have been made. The publisher remains neutral with regard to jurisdictional claims in published maps and institutional affiliations.

Cover pattern © Melisa Hasan

This Palgrave Macmillan imprint is published by the registered company Springer Nature Switzerland AG.
The registered company address is: Gewerbestrasse 11, 6330 Cham, Switzerland

For Diana and Nimrod

Acknowledgments

Writing this monograph as a policy and economic historian, an "outsider" to the group that usually engages in wildlife policy issues, I have been guided by the words of longtime New York State wildlife manager Nate Dickinson, who argued that "clues to the worth of a paper … may be provided by the references identified and the degree to which they were incorporated. … The true professional will be willing to recognize any work that makes a contribution and is technically sound" (1993, 44).

Due to my outsider status, this monograph could not have been written without the generous assistance of the Property and Environment Research Center (PERC) in Bozeman, Montana, and its 2020 Julian Simon Fellowship program. And it would have been far less worthy were it not for the comments and suggestions offered up by PERC's staff and fellows, including Monique Dutkowsky, Randy Rucker, Paul Schwennensen, Brian Yablonski, and especially Wally Thurman, who suffered through two full readings. Finally, Michael 't Sas-Rolfes provided a crucial international context.

Meagan Simpson also deserves my thanks for stepping up in her new role at Palgrave and procuring an excellent anonymous referee who helped me to put the final touches on the manuscript and avoid several serious mistakes. Any and all remaining errors of fact and interpretation, commission, or omission, however, remain the sole responsibility of the author and the SARS-COV-2 virus, the pathogen that causes COVID-19, and, most especially, the policy responses thereto, which destroyed many a well-laid research plan and provided a seemingly endless source of distractions.

CONTENTS

LIST OF FIGURES

)

CHAPTER 1

Introduction: The Problem of Policy Persistence

Abstract Chapter 1: Outlines the chapters of the book in the context of the necessity for successful policies, like the North American Wildlife Conservation Model, to evolve with changing circumstances.

Keywords North American Wildlife Conservation Model • Scientific wildlife management • Policy evolution • Government efficiency and inefficiency

Many parts of government do not function efficiently or even effectively. They cost too much or do not perform the services they were intended to. Law professor Schuck, a Democrat, does an admirable job explaining why (2015).

Although some government policies were flawed from the start, the product of political processes rather than rational policymaking, many worked well at first but did not adapt to changing circumstances, becoming irrelevant or even pernicious by degrees. As years turn into decades, and in some instances centuries, the policies, and the rules, regulations, and budget lines attending them, start to seem like part of the natural landscape, the ecology of the economy. Economies and environments change, but often policies do not evolve, locked in time by statute or bureaucratic inertia. Unlike natural organisms that fail to adapt sufficiently

© The Author(s), under exclusive license to Springer Nature Switzerland AG 2022
R. E. Wright, *The History and Evolution of the North American Wildlife Conservation Model*,
https://doi.org/10.1007/978-3-031-06163-9_1

quickly to changing environments and go extinct, though, concretized policies remain very much in existence, weighing down constituencies they once helped (Wright and Zeiler 2014).

"One of the problems in understanding modern life," economist Julian Simon explained in his final book, "is the implicit comparisons with a past that never existed" (1999, 103). In many cases, origin mythologies arise that elide important details, nuances, and setbacks in order to tell a clean Whiggish story of progress: things were bad, real bad, but then super court case, law, policymaker, or politician X came along and saved the day with agency, law, model, program, or regulation Y. And we all lived happily ever after, except we rarely do.

Some reforms may come, but only to conserve the original core policy, which often creates rents that concentrated interest groups seek to protect. The Security and Exchange Commission's mandate that institutional investors rely upon the credit ratings of a few credit rating agencies, one of the several related proximate causes of the financial crisis of 2008, represents a prime example (Smith and Wright 2021). So, too, does the North American Wildlife Conservation Model (NAWCM) (Dickinson 1993, 101). The former already helped to foment crisis but time remains to reform the latter before crisis strikes wildlife and the communities that support it (Child 2019, 3).

Chapter 2 of this monograph describes the NAWCM and the difficulties it faces confronting circumstances very different from those that created it, namely the local overabundance of white-tailed deer (*Odocoileus virginianus*), feral hogs (*Sus scrofa*), and other wild game animals. It describes scientific wildlife management "in a simple, logical manner" (Dickinson 1993, 45) and argues that it constitutes the sole necessary cause of sustainable consumptive use of living natural resources. It concludes by showing that scientific wildlife management and markets for wild game meat can serve the interests of wildlife if commercial influences on wildlife managers remain adequately checked (Anderson and Leal 2001, 121; Wright 2022).

Chapter 3 reviews the history of wildlife management in North America, from American Indians through the development of the NAWCM in the late nineteenth century, and surveys the interplay between biology, ecology, markets, range, and population size of a dozen important types of wild animals. The key takeaway is that common pool problems, especially when combined with species-specific characteristics, like herding and flocking behaviors, that render human predation particularly inexpensive endanger wild game populations far more than the scientifically managed commercialization of their flesh does.

Chapter 4 warns that the facile application of "democracy" to wildlife management poses grave risks for wildlife, especially wild game species, by threatening the authority of scientific wildlife management and further eroding the wildlife dollars volunteered by consumptive users.

Chapter 5 suggests several ways that wildlife managers, short of reinstating markets for wild game meat, can increase the number of consumptive users and political support for the sustainable, consumptive use of wildlife. The most important suggestion is for the legalization of proxy hunting, where needed to meet local scientific wildlife management goals, as a way to increase wild game harvests short of full-blown re-commercialization.

The conclusion, Chap. 6, argues that scientific wildlife management, checked by the need to maintain inflows of consumptive user dollars, will protect North America's wildlife throughout the twenty-first century better than meat market bans can.

References

Anderson, Terry L., and Donald R. Leal. 2001. *Free Market Environmentalism*, Rev. ed. New York: Palgrave.

Child, Brian. 2019. *Sustainable Governance of Wildlife and Community-Based Natural Resource Management: From Economic Principles to Practical Governance*. New York: Routledge.

Dickinson, Nate. 1993. *Common Sense Wildlife Management: Discourses on Personal Experiences*. Altamont, NY: Settle Hill Publishing.

Schuck, Peter. 2015. *Why Government Fails So Often: And How It Can Do Better*. Princeton: Princeton University Press.

Simon, Julian. 1999. *Hoodwinking the Nation*. New Brunswick: Transaction Publishers.

Smith, Andrew, and Robert E. Wright. 2021. Sowing the Seeds of a Future Crisis: The SEC and the Emergence of the Nationally Recognized Statistical Rating Organization (NRSRO) Category, 1971–75. *Business History Review* 95 (4): 739–64. https://doi.org/10.1017/S0007680521000106.

Wright, Robert E. 2022. The Political Economy of Modern Wildlife Management: How Commercialization Could Reduce Game Overabundance. *Independent Review* 26 (4): 512–32.

Wright, Robert E., and Thomas Zeiler, eds. 2014. *Guide to U.S. Economic Policy*. Washington: CQ Press.

The North American Wildlife Conservation Model and Its Discontents

Abstract Chapter 2: Describes the North American Wildlife Conservation Model and the difficulties it faces confronting circumstances very different from those that created it, namely the local overabundance of white-tailed deer (*Odocoileus virginianus*), feral hogs (*Sus scrofa*), and other wild game animals. It describes scientific wildlife management in a straightforward manner and argues that it constitutes the sole necessary cause of sustainable consumptive use of living natural resources. It concludes by showing that scientific wildlife management and markets for wild game meat can serve the interests of wildlife if commercial influences on wildlife managers remain adequately checked.

Keywords North American Wildlife Conservation Model • Scientific wildlife management • Hunting • Wild game meat markets

The NAWCM is an important policy model generally credited with saving many wild game species from extinction. During its century plus reign, no big game species has gone extinct and several have attained, or approach, record populations, insofar as they can be estimated. By all accounts (see, e.g., Prukop and Regan 2005, 374; Organ et al. 2012, viii–ix, 2; 2016,

© The Author(s), under exclusive license to Springer Nature
Switzerland AG 2022
R. E. Wright, *The History and Evolution of the North American Wildlife Conservation Model*,
https://doi.org/10.1007/978-3-031-06163-9_2

10–12; Feldpausch-Parker et al. 2017, 33–35; Organ 2018, 125–27), seven main components, the so-called seven sisters, comprise the NAWCM's core provisions:

1. The state holds wildlife as a trustee for the people, a.k.a. the public trust doctrine.
2. Markets for wildlife, especially game meat, must be banned or heavily restricted.
3. Allocation of surplus wildlife occurs according to law.
4. Wildlife can be killed only for legitimate purposes, like food or fur.
5. Wildlife is an international resource, so the management of migratory waterfowl, mammals, and marine life must be coordinated internationally.
6. Science, not politics, should drive wildlife management decisions (scientific wildlife management).
7. Hunting must be democratic.

While some consider it the linchpin of the NAWCM (Smith 2011; Organ 2018), the public trust doctrine, which came out of the 1896 US Supreme Court case *Geer v. Connecticut* (Lueck 1995, 4), is broadly misunderstood to signify government "ownership" (a tricky concept indeed, cf. Warnock 2015) of wildlife when in fact it is supposed to mean only trusteeship or stewardship (Huffman 1995, 28). The government, in other words, must act to conserve all wildlife, not just wild game species, for all people, not just hunters, fishers, and trappers (consumptive users), by the best available means, but wildlife are not listed as assets on its balance sheet. Statutes often assert state "ownership" of wildlife, but only to facilitate scientific wildlife management (Braverman 2015, 155–56). Moreover, the public trust doctrine is relatively weak in Canada, which inherited the notion of Crown ownership of wildlife and hence must rely on provincial statute for clarification and authority (Peterson et al. 2016, 429; Kessler 2018, 72–74).

The public trust doctrine also raises the specter of agency costs (Anderson 1998, 279) and questions concerning which parts of the government are responsible for maintaining the public trust doctrine and, most importantly, the checks preventing government officials from using wildlife, or the manipulation of game laws, for private emolument. Smith (2011, 1539–41) implies that government wildlife managers possess incentives to check venal legislators and vice versa, and that other parts of the executive and judicial branches have similar incentives, as do other levels of government, for example, county, provincial, or federal. But the same could be said of the Bureau of Indian Affairs and Indian trusts and that has not turned out so well for American Indians, who if they live on Reservations have been citizens in name only since 1924 (Anderson 1995).

The main checks against malfeasance, in fact, appear to be wildlife commissions, consumptive users, and voters (Barboza and Tihanyi 2018, 381–83). Voters are usually rationally ignorant of the policies of politicians (Caplan 2007) but they can be stirred to oust incumbents caught expropriating public resources. Federalism and the de facto existence of a second wildlife conservation model based on private property also serves to check those who might try to enrich themselves at the expense of wildlife or their consumptive users.[1]

That the NAWCM, as currently understood, might be biased or flawed in some ways will come as no surprise to those conversant with environmental scholarship and public perceptions of it. As Simon (1999) showed, many scientists, and most members of the public, mistakenly believe that environmental conditions always worsen and that population growth and market forces are to blame. False bad news prevails because of prevailing incentives: researchers who focus on crises are more likely to receive funding; journalists pitching horror stories are more likely to gain editorial attention; people are more likely to read and remember warnings of dire outcomes than cautious statements based on nuanced study.

Moreover, some people who believe deeply, if irrationally, in a cause are willing to stretch the truth to achieve their goals because they believe that the ends justify the means (Simon 1999). Overspecialization can also cause policy problems. As Jon Rodiek noted some three decades ago, "progress in resource management is closely related to the continuous development of its separate disciplines" (1991, 3), but also on effective communication between the practitioners of those disciplines. When major texts (e.g., Leopold et al. 2018) do not cover basic relevant principles, like the common pool economics of Ostrom (1990, 1992), unnecessarily suboptimal policies may arise and persist.

[1] Such private property might be better termed non-government as it might be owned communally by an Indian tribe or other communal group, as in Namibia (Barnes et al. 2002). Any ownership structure that promotes stewardship can serve as an alternative to government control via the public trust doctrine.

SCIENTIFIC WILDLIFE MANAGEMENT, THE SAVIOR
OF AMERICA'S WILD GAME

Of the seven main components of the NAWCM, only scientific wildlife management is a necessary cause (Vercauteren et al. 2011, 190–91) of wildlife conservation, a.k.a. sustainable use (Decker et al. 2017). Scientific wildlife managers "must clearly state the problems, form hypotheses, observe and experiment, interpret the data, and then draw logical conclusions" (Dickinson 1993, 7).

Scientific wildlife management is the only of the "seven sisters" that *must* be in place to ensure the continued existence of major mammalian and avian game species. It is not, however, a sufficient cause of conservation in some contexts (Ostrom 1992). In other words, the other six "sisters" can be helpful under some circumstances and can be thought of as tools that wildlife managers can impose or relax to fine tune scientific wildlife management in the face of climatic, ecological, economic, policy, or other shocks (Bosselmann 2015, 6–7).

The entire earth is essentially under human management (Peterson et al. 2016, 431; Couzens et al. 2017, 23–24), some of it passive, some of it active but arbitrary or non-scientific, and some of it active and under scientific principles managed by government and/or private experts. (If you think humans ought not manage ecosystems, read Budiansky 1992, who notes that humans are an integral part of most ecosystems and hence cannot help influencing them in myriad ways great and small.) Passive management entails monitoring but ultimately means leaving nature to itself and is best employed where human interference with natural processes is lowest, as in Antarctica and the ocean depths. Non-scientific management refers to areas that humans exploit as a common resource or where arbitrary rules apply to wildlife resource exploitation, as in parts of Africa, Asia, and South America.

Scientific wildlife management is one of the seven sisters of the NAWCM if the wildlife managers are government workers but outside the model, as it is currently understood, if they are private. In fact, parallel systems of government and private wildlife managers provide the best protection for wild creatures by increasing the likelihood that wildlife management remains on a scientific footing (Lueck 1995, 5). Private wildlife managers "must adhere to the laws" but otherwise have more discretion to innovate than their public sector peers (Dickinson 1993, 8; Huffman 1995, 34), whether they manage wildlife on land primarily used for other purposes, like farming or ranching, or whether they manage private land held primarily as conservation property (Gooden and t' Sas-Rolfes 2020).

In the NAWCM, landowners and riparian rights owners control access to game, as they traditionally have (Anon. 1897, 96; 1909, 305–6; Reid and Nsoh 2016, 10–12), but management generally falls into the hands of government wildlife managers. In some states, however, landowners may assert management rights over wild animals if they erect high fences that restrict animal movement onto and off their lands (Prukop and Regan 2005, 375). They essentially become ranchers of wild game (Lueck 1995, 5) but cannot directly harvest and sell the meat (except that of ungulates like bison [*Bison bison*] where regulated like cattle), instead leasing access to the land and game to hunters and/or selling them a service referred to as "guiding" or "outfitting."

By all accounts, private wildlife managers have better incentives and information to manage their own wildlife populations than bureaucratized public wildlife managers do (Baden 1998; Bish 1998). Private efforts may be small and segmented. Local consortia of landowners, however, can successfully manage populations of non-migratory game (e.g. white-tailed deer or feral hogs), even in the absence of high fences, if contracting costs are low and they understand scientific wildlife management principles (Lueck 1995, 8–11; Anderson 1998, 267–68). Private game preserves so big that game animal behavior is unaffected by fences are also possible (Decker et al. 2017, 825) and date in America to the late nineteenth century (Crossways 1896) when "millions of acres," a million in New York's Adirondack Mountains alone, were under the private control of "able game guardians" (Anon. 1895b.). The entirety of Jekyll Island off the coast of Georgia, most infamous for being the location where millionaires hashed out a plan that led to the formation of the Federal Reserve, was a giant game preserve where fish, fowl, hogs, and deer abounded for the sporting pleasure of men with last names like Field, Morgan, and Vanderbilt (Anon. 1895b; Shaw 2019).

If not management by government wildlife biologists, what, exactly, does scientific wildlife management entail? To be scientific means to test hypotheses, ideas about the way the world works that can be falsified with empirical observation. Strictly speaking, scientists never "prove" a hypothesis or theory, they simply have not yet found enough evidence to prefer some alternative explanation. To be "unable to reject the null" means that a set of data or other empirical observables does not overturn the tested hypothesis, thus confirming but not conclusively proving it because other data, or the same data looked at in a different way, may be found inconsistent with the hypothesis. But science in practice is much messier than any such simple description of its methods (Feyerabend 2011). The history of science is replete with examples of scientific theories that have been tested, tentatively accepted, but later discarded due to the accumulation of countervailing evidence until a crisis necessitates the formulation of a new hypothesis or theory that encompasses more observations than the old one. That is how, and why, science "works" (Kuhn 1996). Scientific wildlife management entails building empirical models of different game species, including population estimates and data on reproduction and sexual maturation rates, longevity, mortality and morbidity rates, habitat requirements, resource constraints like winter feed, and so forth. Ideally, wildlife managers adjust harvest quotas as conditions change and their understanding of species biology improves, increasing quotas and relaxing regulations, for example, when they learned that black bears (*Ursus americanus*) were more fecund in some habitats than previously believed (Hriestienko and McDonald 2007, 78).

Because science is far from "ironclad fact," it can be manipulated, consciously or unconsciously. Indeed, even peer-reviewed science in important medical fields is often flawed (see, e.g., Ioannidis 2005; Formaini 1990). That is why President Dwight D. Eisenhower, in his famous farewell speech warning Americans of the "military industrial complex" also warned that "public policy could itself become the captive of a scientific-technological elite" (1961).

The best science emerges where pluralism prevails (Pennington 2011, 235, 248; Cat 2017) and scientists compete for grant dollars or other emoluments by critiquing each other's work and running experiments and other tests designed to falsify the other scientist's or lab's hypothesis. Independent replication of results is a cornerstone of the scientific method (Malloy 2001, 66–67, 185–86), which is why Nobel physicist Richard Feynman once said that "science is the belief in the ignorance of experts" (1969). Trust but verify and maintain checks, like common sense and professionalism (Dickinson 1993, 6, 20–21), against the imposition of potentially harmful policies. Without professionalism, one long-time wildlife manager claimed, wildlife conservation "programs will necessarily flounder in a mass of fuzz" (Dickinson 1993, 8). Wildlife managers, he insisted, "must be held accountable" so they suffer negative consequences if they implement policies too rashly (Dickinson 1993, 55, 95).

Science alone cannot make proper policy paths clear because scientific studies are subject to differing interpretations and may be flawed in myriad ways (Graham et al. 1988, 179–219). Ergo, wildlife science needs to be undertaken not just by government wildlife biologists, or PETA, or private wildlife managers, think tanks, or consumptive users, but by all such groups (Reid and Nsoh 2016, 16). State wildlife managers, after all, suffer from the same information problem that all central planners do, so no matter how expert or well-meaning they may be, they cannot be infallible (Hayek 1945). Many scientific wildlife management techniques are relatively simple, like tracking the health of habitats and harvested animals, but not always sufficient (Dickinson 1993, 39–43). To prevent serious errors that could lead to catastrophic biological or fiscal results (Dickinson 1993, 69–71), the creation of competing scientific views must be encouraged and the policies most consonant with available data adopted (Fryxell et al. 1991).

Ostrom (1990, 8–22) points out that "Leviathan-only" and "privatization-only" approaches to common pool problems are likely to fall short in particular instances to the extent that they are forcibly imposed upon stakeholders. She suggested allowing those with substantial material interests, for example, wildlife managers and consumptive users, to work out solutions. Discerning the best options means allowing heterogeneity and time for selection processes to work and accepting the result, even if it does not fit some ideological prior.

A monolithic model like the NAWCM is more ideological, an article of faith, than scientific, a hypothesis subject to ongoing real-world testing, and that is dangerous. Unless scientific wildlife management is made paramount, the parent or eldest of the other six siblings in a sense, one or more of the other components of the NAWCM, especially the public trust doctrine or "democracy," can subvert it. Bribes, rational ignorance, vote swapping, and outright rent-seeking can ignore or twist the best science (Shrader-Frechette 1991, 169–96) and poorly done "junk" science can frighten voters and policymakers into taking extreme, unwarranted measures. Banning DDT after publication of Rachel Carson's *Silent Spring*, which erroneously claimed that the pesticide would soon lead to the extinction of many birds and cancer in most humans, represents one of many examples (Malloy 2001, 145–46; Meiners et al. 2012).

Unlike proponents of the NAWCM, scientific wildlife managers would not make categorical statements like "wild game meat sales must always be banned" but rather would establish the conditions under which wildlife parts (antlers, bones, claws, feathers, hides, teeth [New York Times 1877; Shields 1887; Forest and Stream 1894; Anon. 1895a]) and especially wild meat markets should be banned, regulated, or even encouraged. Historical (change over time) and comparative (change over place) studies can help to create such parameters. *The key to successful policy evolution is ensuring that decision-makers have incentives to find and analyze real world feedback and make appropriate adjustments when necessary, on the basis of reason, not ideology* (Child 2019, 7).

Clearly, when game populations are low and declining, commercial harvest creates poor incentives while sport hunting can generate political pressures and revenues that can lead to conservation measures. When game populations are high and growing and sport hunters cannot harvest enough to prevent overabundance, however, commercial harvest, under scientific wildlife management restrictions, could make up the deficit at lower cost than relying on expensive culls or automobile collisions (Vercauteren et al. 2011, 186–87).

Wildlife overabundance constitutes a serious problem in some areas. Deer cause over $170 million of crop damage per year, plus millions more dollars of damage to landscaping and gardens. Animal spread or vectored diseases like Lyme's, crop predation, and undesirable human-animal (e.g., backyard bears and beaver dams) encounters can lead to everything from property damage to wild animals becoming habituated to humans, which is rarely a good thing (Dickinson 1993, 28–30, 67; Jonker et al. 2006, 1009–10; Hriestienko and McDonald 2007, 72–73; Vercauteren et al. 2011, 186–87).

Economists can play a key role in determining commercial hunting license pricing. If licenses are priced too high, not enough game will be commercially harvested to meet scientific wildlife management goals. If commercial licenses are too cheap and flat fee (as opposed to per tag or per harvested animal), they can encourage over-harvesting, as Missouri discovered in the 1870s when several counties implemented $25 flat fees for commercial or "pot hunters" as they were known. As a commercial hunter named Col. Pratt explained, he paid the $25 but made sure that he "soon got more than the value of it in game" (Anon. 1876).

Scientific wildlife management was the necessary cause of the reversal of the downward trend of wild game numbers (though not always range [Laliberte and Ripple 2004], especially marginal range [Dickinson 1993, 80–82]) in North America at the end of the nineteenth century and the beginning of the twentieth. It was not *then* a sufficient cause, however, due to the peculiar context of the times. The other six sisters, particularly banning wild meat markets and articulation of the public trust doctrine, combined with propitious external factors to create conditions sufficient to reverse the decline in game populations.

State laws that effectively banned the sale of wild game meat, aided by the federal Lacey Act of 1900 which forbid the interstate sale of illegally harvested wildlife (Vercauteren et al. 2011, 185), made it easier and cheaper for wildlife managers to implement "bag and tag" restrictions by reducing incentives to poach and by inducing fancy hotel, restaurant, and grocer suppliers to switch to domesticated varieties (Anon. 1898a), like farmed ducks instead of wild mallards (*Anas platyrhynchos*) and ranched reindeer instead of wild caribou (*Rangifer tarandus*). Restrictions on interstate shipments also raised the price of wild game relative to farmed substitutes (Anon. 1899c). Similar enactments in Canada followed (Organ et al. 2012, 5).

Earlier state restrictions on the sale of game meat outside of open season helped pave the way for the Lacey Act (Anon. 1874). Locking up poachers and illegal shippers and hitting them with stiff fines and equipment forfeiture (Anon. 1887) helped (Anon. 1892, 1895c, 1895d, 1896b, 1896c, 1898c, 1899d), though some inevitably came up with ingenious ways of hiding illegal game shipments, like shipping dead deer in human coffins (Anon. 1891). Imposition of onerous regulations, short of outright bans, on in-season, intra-state venison markets also helped to end wild meat markets (Anon. 1899e).

Reducing poaching by market hunters made management laws an easier sell with sportsmen, who no longer could blame "pot hunters" for shooting "all" the game (O. 1889). Within a generation, wild game switched from being a delicacy for the well-heeled (see Fig. 2.1) into something that poor people ate out of necessity or sportsmen consumed out of tradition or taste preferences. Today, when people buy venison in

PARTRIDGES
AND
VENISON
AT
LYNCH'S

Fig. 2.1 Wild game advertised by fancy restaurant, 1898. (Redrawn from original source: Anon. 1898d)

one of North America's many fine restaurants, or Arby's, they are not buying the perfectly seared flesh (or utterly destroyed flesh in the case of Arby's) of a wild white-tailed deer but rather another species of cervid, red deer (*Cervus elaphus*), ranched most likely in New Zealand.

Articulation of the public trust doctrine, along with notions of hunter democracy and internationalism, made it clear that everybody had an interest in protecting wild game with the aid of government wildlife managers and their enforcers, who traditionally were rightly called game *wardens* (Anon. 1895c, 1896b, 1896c), that is, people "responsible for the supervision of a particular place or thing or for ensuring that regulations associated with it are obeyed." What those tenets of the NAWCM did was to break the traditional notion that animals were God-given and hence owned by humankind in common. They thus reduced the "tragedy of the commons," or the incentive to harvest resources before somebody else does, which can lead to resource exploitation rates that exceed the natural replacement rate (Anderson 1998, 259; Peterson et al. 2016, 430).

As Ostrom (1990) showed, common pool resources can sometimes be shared in a sustainable manner. The tragedy of the commons can occur without commercialization of the resource, but the rate of degradation will increase the larger the market for the resource because the incentive to harvest it before others do will be correspondingly increased (Johnston et al. 1896). Markets are problematic because they divorce incentives to harvest from local human population levels. When flesh is not sold, whether due to legal, economic, physical, or cultural barriers, some may be preserved for later consumption but, because humans prefer dietary variety and preservation methods can delay consumption only for weeks, months, or at most a few years, depending on available preservation technologies, local hunting pressure ultimately is limited by the size of the local population.

In the eighteenth century in the Churchill River area, for example, residents killed only as many geese (species unclear) as they could "consume fresh" during the fall migration, which lasted only a few weeks, and in the spring they killed and salted only as many geese as they needed for the summer (Wales 1770, 121–23, 126). If those hunters had tried to supply the demand for geese in Philadelphia or London, however, they might have killed geese until they ran out of ammunition, or the birds themselves, if the market price was high enough, calculating that if they did not harvest the birds, somebody else, like the Cree (Richards 2014, 31), would.

In other words, markets induced a flow of game animals from places where they were plentiful relative to humans to urban markets. In 1880, for example, two market hunters in Colorado harvested about 500 prong-horn (*Antilocapra americana*), the meat of which sold in Denver for 7 or 8 cents a pound, and 250 wapiti (*Cervus canadensis*), which sold from 7 to 10 cents per pound. Deer, which might have included mule deer (*Odocoileus hemionus*) in addition to white-tailed deer, sold in Denver at 10 to 12.5 cents per pound. Hides provided additional revenue (Fort Collins Express 1881).

To combat the tragedy of the commons, the public trust doctrine turned ownership/stewardship of wild animals over to the government, which gave it a *credible incentive to conserve the resource* via investment in professional wardens and other enforcement mechanisms (Johnston et al. 1896), thus reducing panicked harvesting in much the same way that deposit insurance stops bank runs (Wright 2010). The public trust doc-trine was also essential in inducing hunters to accept licensing systems that were still so novel around the turn of the twentieth century that they had to be described in newspapers (Anon. 1899a).

The international tenet of the NAWCM was necessary so that hunters would not see migratory animals as subject to the commons problem. While the defense of Wisconsin's *spring* duck season on the grounds that other states also allowed spring hunts was eventually met with federal reg-ulation (Kreppel 1897; Anderson 1998, 278), only international regula-tion of wildfowl, like the 1916 migratory game bird treaty with Canada/the UK (Lueck 1995, 4), could completely end the common pool problem.

Positive exogenous shocks, including the reduced range of natural predators and habitat renewal, especially afforestation, reforestation, and the creation of "urban forests" (Braverman 2015, 159–60), also helped the NAWCM to reverse game animal population decline. Cheaper farmed protein and fur farming helped too. Due to the NAWCM and those exter-nal factors, populations of black bear, deer, feral hogs, turkeys (*Meleagris gallopavo*), snow geese (*Chen caerulescens*), and some other game animals have grown too quickly in some areas. Tag and bag limits and other hunt-ing regulations were too stringent in those cases, reducing human welfare (the positive emotions some associate with hunting, harvesting, and con-suming wild game) below maximum sustainable levels, but most would agree it was better to err on the side of caution and harvest too few, rather than too many, game animals.

What has become clear is that in most areas, human predation from hunting and vehicular accidents is now the main check on the number of white-tailed deer (Grovenburg et al. 2011) and other important game species, including other ungulates and turkeys. Vehicle collisions kill approximately 100,000 deer a year in New York State alone (Braverman 2015, 172) and 700,000 to over a million nationwide. In addition to threatening human life, such collisions cost more than $390 million, making them a major problem rather than a sustainable management tool (Schwabe and Schuhmann 2002, 609).

For complex reasons, the number of hunters has declined, even in largely rural states with strong hunting traditions like Montana (Schorr et al. 2014, 944) and Wisconsin (Winkler and Warnke 2013, 460–61). Today, hunters in many areas harvest too few animals even after "bag and tag" limits and other hunting regulations, like season lengths and acceptable hunting equipment and techniques, are relaxed. Harvest rates simply have not kept pace with the continued secular decline in the number of active hunters, which has had the secondary effect of rendering some wild animals less fearful of humans than politically and socially desirable (Hriestienko and McDonald 2007, 79).

Clearly, reforms are needed but the rigidity of the NAWCM keeps the most attractive alternatives off limits. Meanwhile, alternatives thought to be in accordance with the NAWCM have encountered considerable difficulties.

Reintroducing natural predators could help to manage game numbers but ranchers fear livestock predation and programs to compensate them (e.g., Anderson and Leal 2001, 172–73; Fischer 2001), long advocated by the Property and Environment Research Center (PERC), have encountered practical difficulties like accurately confirming kills (Dickie 2018), just as Anderson (1998, 266) warned. Finding funding is also fraught because the beneficiaries of reintroduction are diffuse (Reid and Nsoh 2016, 3). Moreover, suggestions that wolf (*Canis lupus*) packs should be reintroduced into urban and suburban America are biologically unrealistic. Wolves have been known to prey on humans but for the most part eschew the heavily populated areas where their deer culling services would be most valuable (Savage 1825, 36; Braverman 2015, 170).

Other approaches are economically impractical. Sterilizing wild animals is wildly expensive (Hriestienko and McDonald 2007, 83; Braverman 2015, 163–69), as is culling them by a professional sniper or professional beaver/rodent exterminator (Jonker et al. 2006, 1019; Lebel et al. 2012,

1431). Sharpshooting over bait worked to reduce deer densities and vehicular collisions in suburban Minneapolis but cost $100 to $200 per cull, which can entail many thousands of dollars of expenditures annually to keep populations near management targets (Doerr et al. 2001).

Non-lethal deterrents or translocation can help to protect specific assets or resources but of course just push the problem down the road, literally or temporally, because they do not address the core overpopulation problem (Vercauteren et al. 2011, 187).

So far, reasoned calls for wildlife managers to allow the commercial harvest of game mammals in areas where they are overabundant have fallen flat because of the prevailing conceptualization of the NAWCM (Prukop and Regan 2005, 375; Lebel et al. 2012, 1431). The sale of wild game meat in North America has become a repugnant market, akin to the sale of kidneys and other human organs, because commercialization is closely associated with slavery and extinction (Freese 1997, 1–5; Peterson et al. 2016, 430; Reid and Nsoh 2016, 24). Once people reject the morality of the sale of some object or service, inducing them to see the benefits of unregulated exchange, or the costs of prohibiting such exchange, becomes extremely difficult (Roth 2015). Hunting horses and consuming horse meat is so repugnant in North America, for example, that wild herds are now managed by giving people up to $1000 to adopt one (https://www.blm.gov/programs/wild-horse-and-burro/adoptions-and-sales/adoption-incentive-program)!

Short-term, therefore, reformers must think in terms of "second-best world" options that improve outcomes but remain shy of optimal policy. One such option, hitherto unconsidered, is to urge wildlife managers to allow designated or "proxy" hunters to sell the service of harvesting wild game instead of selling the meat itself. Under this proposal, detailed in Section 5, proxy hunters lawfully could charge clients a fee for harvesting and processing wild game on their behalf. "Tag and bag" and other regulations would still apply and the retail sale of game meat by weight would still be prohibited but limited monetary incentives would be created for avid, skilled hunters to provide wild game meat to those too old, squeamish, or busy to harvest it themselves.

Longer-term reformers must confront the reasons why people find the free exchange of specific goods to be repugnant. With human organs, complete liberalization may never be achieved due to deep-seated cultural and religious views. Repugnance of the sale of game meat, by contrast, can be eliminated in time by recasting the NAWCM's origin story in more realistic terms, by comparing its outcomes with those of countries that achieved conservation goals without relying on all of the NAWCM's core tenets, and by highlighting the conservation success of reforms like proxy hunting.

THE COEXISTENCE OF WILD MEAT MARKETS AND SUCCESSFUL SCIENTIFIC WILDLIFE MANAGEMENT

Policymakers should not allow "principles" based on an incomplete under-standing of economics to "become encoded as truth" (Harris 1984, 7). Banning wild meat markets is not a necessary condition of scientific wild-life management success. For starters, commercialization of a species is not a sufficient cause of its demise. Most tellingly, markets for wild game flour-ished in England after it privatized game (Lueck 1995, 19). In Sweden and many other European countries, the retail sale of big game meat has also been allowed without adversely affecting wildlife populations (Ljung et al. 2012, 669).

South Africa and other nations in the southern part of the African con-tinent have also long allowed the commercial sale of wild game, which sparked a rebound in wild game populations as ranchers learned that wild animals did not so much compete with cattle, sheep, and goats as they complemented them. Many ranchers now diversify their businesses by rais-ing domestic livestock, providing ecotours, harvesting wild animals to sell their meat, and guiding trophy hunters. In 2003, South Africa alone sup-ported about 5000 game ranches and 4000 mixed domestic-wild ranches. Wildlife there, like on many private Texas ranches working under similar institutions, has achieved numbers not seen in many decades. In both Texas and South Africa, wildlife thrive because they "pay their own way" through their commercial sale (Carruthers 2008). But only in Africa can ranchers benefit from the commercial sale of their flesh.

Even within North America some wildlife species remain unprotected by the NAWCM and yet they have not been exterminated (Braverman 2015, 157). Sometimes, even leveraging market incentives through bounty programs that create government-funded "markets" for the scalps or tails of so-called nuisance animals failed when the bounties proved insufficient to cover costs, including the opportunity cost of the marginal hunter's or trapper's time (Anon. 1842, 345; 1898b). In fact, concerted strenuous efforts to eradicate some species, like prairie dogs, with poison, fire, and even steam failed (Anon. 1858, 233).

Moreover, efforts to extirpate a species through bounties often spur unintended consequences. Trappers eager to earn bear bounties, for instance, slaughtered deer simply to keep their traps freshly baited (Glenwood Daily Avalanche 1891). In New England, bear trappers engaged in arbitrage by dragging bears from Maine, where the bounty was $5, to New Hampshire, where it was $10, before dispatching them (Lewiston Journal 1892). Today, many furry victims of vehicles in Minnesota, Iowa, and Nebraska end up in South Dakota, where nest predator tails receive $10 from the state, as evidenced by the many tail-less badgers (*Meles meles*), raccoons (*Procyon lotor*), red fox (*Vulpes vulpes*), opossums (*Didelphis virginiana*), and skunks (*Mephitis mephitis*) visible along highways in those states.

Pack predators like wolves were pushed out of agricultural areas through a combination of bounties, professional hunters paid by stock associations (Anon. 1899), large-scale winter round-up hunts (St. Louis Globe-Democrat 1895), trapping, and poison (Forest and Stream 1896). Poisons could backfire though as some, like strychnine, were thought to cause post-mortem fur loss, which rendered them worthless (Anon. 1871). Poisoning also trained wolves to avoid dead bait, inducing wolf hunters and trappers to begin using old horses as live bait (Detroit Free Press 1872).

Obviously, living creatures harvested only after they have reproduced and died can be commercially exploited without much fear of causing extinction. For example, large markets for Spanish moss, which locals used to stuff mattresses and chair bottoms and to wad their guns, existed in the nineteenth-century South. Because the moss was gathered only after it died, however, it not only survived its commercialization, but also covered "whole forests" (Davis 1817, 27–29, 37, 43). Similarly, removing living but reproductively dead animals from populations will generally help the remaining animals by freeing up resources for breeders.

Some ocean fisheries subject to the commons problem and commercialization, especially in ecologically sensitive areas like the polar regions, have suffered collapse (Couzens et al. 2017; Liu et al. 2019), but others claimed by just a few countries have remained stable (Honneland 2013). Like game animals, fish (and sharks: https://www.fisheries.noaa.gov/insight/understanding-atlantic-shark-fishing) have been scientifically managed in many of North America's fresh and coastal saline waters and in many areas thrive despite the inherent uncertainty of ascertaining fish numbers, especially in the ocean (Biber and Eagle 2015, 787–88, 803). Markets for "catch shares" largely have proven successful (Huggins 2013,

67–69), as have other "bottom up" commons management solutions (Leal 1998, 284).

Overstressed commercial fisheries were the result not of markets per se but of the ability of commercial fishers (and whalers, cf. Peterson 1992) to induce wildlife managers to relax scientific management rules (Biber and Eagle 2015, 806–7, 826–28). Captive cervid ranchers have also been able to pressure wildlife managers or, as in North Carolina, to change regulators entirely (Brown 2016, 22–23). Economists call such behaviors rent-seeking and stress that they are essentially the antithesis of market competition (Anderson and Hill 1989).

Moreover, some purely recreational fisheries have been stressed by poaching due to the difficulty of monitoring the activities of individual fishers (Biber and Eagle 2015, 832). Hunting dogs with deer was discouraged by making it legal for deer hunters to shoot any dog in the act of chasing deer (Anon. 1894; Dickinson 1993, 94), but no parallel exists to reduce incentives to poach fish.

Wildlife managers who would never consider commercial hunting regularly keep fish numbers in check by licensing commercial fishers, most of whom use netting or hook-and-line techniques that allow them to target specific species without unduly stressing game fish (see, e.g., South Carolina's regulations: http://www.eregulations.com/southcarolina/huntingandfishing/nongame-methods-devices/). Hunting is not generally considered "catch and release," but if conducted in daylight, it can be as selective as catch-and-release fishing because hunters can decline to harvest animals that do not meet cull criteria (Wright 1868b, 471; Miles 1895, 485; Waselkov 1978, 18–19).

Interestingly, alligators (*Alligator mississippiensis*) are commercially harvested in Louisiana after it successfully transitioned from an open access to a scientific wildlife management regime in the 1970s (Joanen et al. 1997, 466). Recent years have seen harvests of 35,000 wild gators out of an estimated population of 1.5 million (https://www.louisianaalligators.com/alligator-management-program.html), up from harvests of 25,000 out of a population of about 700,000 in the 1990s (Joanen et al. 1997, 469–71, 480). Gators are reptiles, not mammals, but many of their biological parameters (lifespan, size, maturation rate) roughly match those of black bears and other mammalian apex predators (Seay 2019; Hriestienko and McDonald 2007, 76–78). Unlike bears and cougars (*Puma concolor*), however, their commercial harvest is fairly indiscriminate as to gender and age as gators are not easily released if caught with a baited hook and line, while bears, cougars, and raccoons can be aged, sexed, and passed with some accuracy, especially if baited or treed (Wright 1868a, 124; Whitney 1931, 38). Alligators produce more offspring than mammals of their size, but their young suffer much higher rates of predation.

Overall, then, the successful management of alligators, despite the commercial sale of their flesh and skins, bodes well for the ability of wildlife managers to manage the commercial harvest of game mammals (Joanen et al. 1997, 473–74; Organ et al. 2012, 17–18). Like cervids, alligators are also being ranched, and have been since the late 1980s (Dickerson 1987). If commercial hunting of alligators ends, it will be because of competition from alligator ranching, not the extermination of the species. Louisiana ranches alone hold 600,000 to 900,000 gators (https://www.louisianaalligators.com/alligator-management-program.html).

Similarly, bullfrogs (*Lithobates catesbeianus*) and other amphibians were long commercially hunted without endangering their existence and continue to be to this day (Organ et al. 2012, 15). In terms of their size and markets for their flesh, they are similar to small mammalian game like squirrels and rabbits. Their long-term successful commercial harvest is not conclusive evidence that small mammal flesh could again be sold safely at market, but it is probative.

During the Civil War, two men made $12 to $18 a night frogging in the Montezuma swamps west of Syracuse, New York, spearing 200 to 300 bullfrogs per night, for which they received $6 per 100 in Buffalo and Manhattan (Anon. 1862, 35). In the 1880s, froggers in the swamps near St. Clair Lake in eastern Michigan supplied the Detroit and eastern markets. A pair of hunters with a dozen years' experience could then make $15 a week apiece jigging or shooting between 40 and 200 bullfrogs a day with a 0.410 shotgun. Despite commercial hunting pressure, the frogs numbered in the millions and got as big as nine pounds, with seven pounders not uncommon (Anon. 1881, 197). Philadelphians managed to supply the local market themselves by tapping the Delaware River and adjacent wetlands for frogs that they sold to a local market for between 30 and 50 cents a dozen, depending on their size. By the 1880s, old froggers pined for the days when they could catch 30 to 40 dozen frogs per outing, instead of the 8 to 12 dozen that had become common. The problem was not hunting pressure, the old froggers lamented, but loss of habitat due to pollution (Philadelphia Press 1883).

Where habitat remained hospitable to frogs, they continued to thrive, as in Friendship, New Jersey, where for the better part of the 1890s Miss Mona Selden made $1600 a year frogging, versus the $520 a year she had been making teaching (Anon. 1896a). Unlike most froggers, Selden owned her own 20-acre patch of swamp, which she apparently managed well. Long thought impossible to farm, especially after failed experiments during the Great Depression (Laskow 2017), bullfrogs have been successfully raised for decades in southeastern Idaho (Troyer 2019) and elsewhere (Patera 1978) for food, pet, and scientific markets.

Similarly, turtle trappers used small shotguns to shoot small birds that they used to bait turtle traps. They sold the turtles (probably *Kinosternon*) to French people living in the United States, who paid "good prices" for the "shellbacks" (San Francisco Examiner 1890). Turtle harvest continues to this day, though more for Asian markets than French ones (Organ et al. 2012, 15).

Trappers generally are not interested in eating the meat of the mammals they trap (Palmer 1939, 17), but some eat beaver (*Castor canadensis*) (San Francisco Call 1885; Speck 1915, 293), marten (*Martes americana*) (Anon. 1866), muskrat (*Ondatra zibethicus*), raccoon "when cooked properly" (Anon. 1899b), and even skunk and badger (Palmer 1939, 17; Griffiths and Thomas 1997, 15). Commercial market demand for the meat of most furbearer species, though, would probably be limited to the pet food market at best. The main prize are pelts and furs, which trappers sell into a global market with minimal scientific wildlife management. Regulation typically takes the form of closed or shortened seasons and lawful traps and set techniques (Dickinson 1993, 51). Generally, trappers kill as many furbearers as they can during open season, which generally matches the months when their fur is thickest and most valuable. Market conditions naturally regulate trappers, who move into and out of the business with fur prices, and into and out of areas with the success of their traps.

Few commercial trappers use live traps, cable restraints, or other catch-and-release technologies (Vantassel et al. 2010, 935), but where and the way they place their sets, and the traps they use, target specific species with considerable accuracy. Nobody ever caught a wolf in a raccoon handhold trap and no ermine (*Mustela erminea*) ever ended up in a lynx (*Lynx canadensis*) snare, but sometimes species others than those targeted, like coyotes and birds, get killed in traps (Mighels 1897). Trapping provides perhaps the clearest indication that mammal species can be commercially harvested with minimal management provided the total economic cost of harvesting them remains high compared to their market prices. When traps can be set and retrieved by inexpensive, long-distance drones, stringent tag limits will likely be required.

Extinction and its precursors, like loss of range (Laliberte and Ripple 2004), stem from complex interactions between a given species' biology, environmental pressures (including habitat and natural predators), and human incentives to harvest a given species. Markets affect the final variable, the intensity of human predation, but represent only one factor of several, including the average cost of harvesting biologically optimal culls. Wildlife managers in North America no longer confront much commercial poaching, but they need to keep in mind that illegal, unregulated, underground markets can appear under the right economic circumstances, and in some instances it might be better to legalize, regulate, and tax them rather than expend resources trying to ban them entirely. (Rhinoceros management in Africa provides a compelling example [Child 2012].)

Economists could model the effect of markets, legal or otherwise, on wild game but if mis-specified such models could lead to the adoption of harmful policies. A theoretically informed empirical approach is safer and methodologically *au courant* (Angrist et al. 2017). An economic history of North America's wildlife can establish a solid base for such studies and also clarify the precise reasons underlying the NAWCM's initial success.

REFERENCES

Anderson, Terry L. 1995. *Sovereign Nations or Reservations? Indian Economies: An Economic History of American Indians*. Pacific Research Institute.
———. 1998. Viewing Wildlife Through Coase-Colored Glasses. In *Who Owns the Environment?* ed. Peter J. Hill and Roger E. Meiners, 259–282. New York: Rowman & Littlefield.
Anderson, Terry L., and Peter J. Hill. 1989. *The Birth of a Transfer Society*. New York: University Press of America.
Anderson, Terry L., and Donald R. Leal. 2001. *Free Market Environmentalism*, Rev. ed. New York: Palgrave.
Angrist, Joshua, Pierre Azoulay, Glenn Ellison, Ryan Hill, and Susan Feng Li. 2017. Economic Research Evolves: Fields and Styles. *American Economic Review: Papers & Proceedings* 107 (5): 293–297.
Anon. 1842. Walton and Cotton's Complete Angler.... *North American Review* 55 (117): 343–372.
———. 1858. Steam Squirrel Hunting. *Scientific American* 13 (30): 233.
———. 1862. Frog Hunting. *Scientific American* 7 (3): 35.

———. 1866. The Marten Trappers. *Frank Leslie's Illustrated Newspaper* (13 Oct.), 60.

———. 1871. The Winnipeg Trappers. *Independent Statesman* (23 Nov.), 59.

———. 1874. The Sale of Venison to Be Stopped. *Cleveland Daily Herald* (2 Dec.), 8.

———. 1876. Hunting and Fishing. *St. Louis Globe-Democrat* (209 Oct.), 6.

———. 1881. Michigan Frogs and Frog Hunting. *Scientific American* 45 (13): 197.

———. 1887. Costly Venison. *Daily Picayune* (10 Oct.), 4.

———. 1891. Shipping Venison in Coffins. *Milwaukee Journal* (12 Nov.).

———. 1892. Big Game Seizure. *St. Paul Daily News* (17 Feb.).

———. 1894. Hunting with Dogs. *Milwaukee Journal* (7 Nov.), 5.

———. 1895a. Hunting Egrets in Mexico. *Daily Inter Ocean* (2 Dec.), 3.

———. 1895b. The Hunting Paradises. *Daily Picayune* (29 Sept.), 21.

———. 1895c. Venison Coming In. *Rocky Mountain News* (30 Nov.), 8.

———. 1895d. Game Birds and Venison. *Bismarck Tribune* (31 Dec.).

———. 1896a. A Frog Hunter. *Emporia Daily Gazette* (12 June).

———. 1896b. Loaded with Venison. *Rocky Mountain News* (11 Jan.), 4.

———. 1896c. Tons of Venison. *Bismarck Tribune* (5 Dec.).

———. 1897. Riparian Rights. Navigable Waters. Trespass. Rights of Hunters. Hall v. Alford, 72 N.W. Rep. 137 (Mich.). *Yale Law Journal* 7 (2): 96.

———. 1898a. It Will Stop 'Pot Hunting'. *Milwaukee Journal* (14 Feb.), 6.

———. 1898b. Hunting in Maine. *Atchison Globe* (5 Oct. 1898), 3.

———. 1898c. For Violating Game Law. *Milwaukee Daily Sentinel* (19 Feb.), 10.

———. 1898d. Partridges and Venison at Lynch's. *Bangor Daily Whig and Courier* (17 Nov.), 3.

———. 1899a. License for Hunting. *Milwaukee Journal* (21 July), 8.

———. 1899b. The Hunting Season. *Bangor Daily Whig and Courier* (20 Oct.), 8.

———. 1899c. Little Venison Yet. *Milwaukee Daily Sentinel* (7 Nov.), 5.

———. 1899d. Venison for Boston Seized. *Boston Daily Advertiser* (4 Mar.), 3.

———. 1899e. Will Not Handle Venison. *Bangor Daily Whig and Courier* (16 Sept.), 8.

———. 1909. Interference with Hunting Rights. *Harvard Law Review* 22 (4): 305–306.

———. 1899. Greatest Wolf Hunter. *Atchison Globe* (15 Aug.), 3.

Baden, John A. 1998. A New Primer for the Management of Common-Pool Resources and Public Goods. In *Managing the Commons*, ed. John A. Baden and Douglas S. Noonan, 2nd ed., 51–62. Bloomington: Indiana University Press.

Barboza, Perry S., and Daniel Tihanyi. 2018. State Wildlife Policy in a National Environment. In *North American Wildlife Policy and Law*, ed. Bruce D. Leopold, Winifred B. Kessler, and James L. Cummins. Boone and Crockett Club: Missoula.

Barnes, Jonathan I., James Macgregor, and L. Chris Weaver. 2002. Economic Efficiency and Incentives for Change with Namibia's Community Wildlife Use Initiatives. *World Development* 30 (4): 667–681.

Biber, Eric, and Josh Eagle. 2015. When Does Legal Flexibility Work in Environmental Law? *Ecology Law Quarterly* 42 (4): 787–840.

Bish, Robert L. 1998. Environmental Resource Management: Public or Private? In *Managing the Commons*, ed. John A. Baden and Douglas S. Noonan, 2nd ed., 65–75. Bloomington: Indiana University Press.

Bosselmann, Klaus. 2015. *Earth Governance: Trusteeship of the Global Commons.* New York: Edward Elgar.

Braverman, Irus. 2015. Conservation and Hunting: Till Death Do They Part? A Legal Ethnography of Deer Management. *Journal of Land Use & Environmental Law* 30 (2): 143–199.

Brown, Robert D. 2016. The Politics of Deer-Farming in North Carolina— Lessons Learned. *Wildlife Society Bulletin* 40 (1): 20–24.

Budiansky, Stephen. 1992. *The Covenant of the Wild: Why Animals Chose Domestication.* New York: William Morrow.

Caplan, Bryan. 2007. *The Myth of the Rational Voter: Why Democracies Choose Bad Policies.* Princeton: Princeton University Press.

Carruthers, Jane. 2008. 'Wilding the Farm or Farming the Wild?' The Evolution of Scientific Game Ranching in South Africa from the 1960s to the Present. *Transactions of the Royal Society of South Africa* 63 (2): 160–181.

Cat, Jordi. 2017. The Unity of Science. In Edward N. Zalta, ed. *The Stanford Encyclopedia of Philosophy.* https://plato.stanford.edu/archives/fall2017/entries/scientific-unity/.

Child, Brian. 2012. The Sustainable Use Approach Could Save South Africa's Rhinos. *South Africa Journal of Science* 108 (7–8): 1–4.

———. 2019. *Sustainable Governance of Wildlife and Community-Based Natural Resource Management: From Economic Principles to Practical Governance.* New York: Routledge.

Couzens, Ed, Alexander Paterson, and Sophie Riley. 2017. *Protecting Forest and Marine Biodiversity: The Role of Law.* New York: Edward Elgar.

Crossways, Diana. 1896. Stag Hunting. *Portland Oregonian* (10 May), 15.

Davis, John. 1817. *Personal Adventures and Travels Four Years and a Half in the United States of America.* London: W. McDowall.

Decker, Daniel J., John F. Organ, Ann B. Fortschen, Cynthia A. Jacobson, William F. Siemer, Christian A. Smith, Patrick E. Lederle, and Michael V. Schiavone. 2017. Wildlife Governance in the 21st Century: Will Sustainable Use Endure? *Wildlife Society Bulletin* 41 (4): 821–826.

Detroit Free Press. 1872. Going for Wolves. *Hawaiian Gazette* (27 March).

Dickerson, A.J. 1987. Rising Demand for Meat Spawns New Industry: Alligator Ranching. *Los Angeles Times*, 24 May.

Dickie, Gloria. 2018. When Cattle Go Missing in Wolf Territory, Who Should Pay the Price? *High Country News*. https://www.hcn.org/issues/50.12/wolves-when-cattle-go-missing-in-wolf-territory-who-should-pay-the-price.

Dickinson, Nate. 1993. *Common Sense Wildlife Management: Discourses on Personal Experiences*. Altamont, NY: Settle Hill Publishing.

Doerr, Michelle L., Jay B. Aninch, and Ernie P. Wiggers. 2001. Comparison of 4 Methods to Reduce White-Tailed Deer Abundance in an Urban Community. *Wildlife Society Bulletin* 29 (4): 1105–1113.

Eisenhower, Dwight D. 1961. Farewell Address. https://www.ourdocuments.gov/doc.php?doc=90&page=transcript.

Feldpausch-Parker, Andrea, Israel D. Parker, and Elizabeth S. Vidon. 2017. Privileging Consumptive Use: A Critique of Ideology, Power, and Discourse in the North American Model of Wildlife Conservation. *Conservation and Society* 15 (1): 33–40.

Feyerabend, Paul. 2011. *The Tyranny of Science*. Cambridge: Polity Press.

Feynman, Richard. 1969. What Is Science? *The Physics Teacher* 7 (6): 313–320.

Fischer, Hank. 2001. Who Pays for Wolves? *PERC* 19: 4.

Forest and Stream. 1894. Corner in Elk Teeth. *Milwaukee Daily Sentinel* (28 Oct.), 13.

———. 1896. Trapping in Wisconsin. *Emporia Daily Gazette* (1 Feb.).

Formaini, Robert. 1990. *The Myth of Scientific Public Policy*. New Brunswick, NJ: Transaction Publishers.

Fort Collins Express. 1881. One Season's Hunting. *St. Louis Globe-Democrat* (2 Jan. 1881), 11.

Freese, Curtis H. 1997. The 'Use It or Lose It' Debate: Issues of a Conservation Paradox. In *Harvesting Wild Species: Implications for Biodiversity Conservation*, ed. Curtis H. Freese. Baltimore: Johns Hopkins University Press.

Fryxell, John M., David J.T. Hussell, Anne B. Lambert, and Peter C. Smith. 1991. Time Lags and Population Fluctuations in White-Tailed Deer. *Journal of Wildlife Management* 55 (3): 377–385.

Glenwood Daily Avalanche. 1891. Slaughtering Game. *Rocky Mountain News* (23 Aug.), 12.

Gooden, Jennifer, and Michael 't Sas-Rolfes. 2020. A Review of Critical Perspectives on Private Land Conservation in Academic Literature. *Ambio* 49 (1): 019–034.

Graham, John D., Laura C. Green, and Marc J. Roberts. 1988. *In Search of Safety: Chemical and Cancer Risk*. Cambridge: Harvard University Press.

Griffiths, Huw I., and David H. Thomas. 1997. *The Conservation and Management of the European Badger (Meles meles)*. Strasbourg: Council of Europe.

Grovenburg, Troy W., Christopher C. Swanson, Christopher N. Jacques, Christopher S. Deperno, Robert W. Klaver, and Jonathan A. Jenks. 2011. Female White-Tailed Deer Survival Across Ecoregions in Minnesota and South Dakota. *American Midland Naturalist* 165 (2): 426–435.

Harris, Larry D. 1984. *The Fragmented Forest: Island Biogeography Theory and the Preservation of Biotic Diversity*. Chicago: University of Chicago Press.

Hayek, Friedrich. 1945. The Use of Knowledge in Society. *American Economic Review* 35 (4): 519–530.

Honneland, Geir. 2013. *Making Fishery Agreements Work: Post-Agreement Bargaining in the Barents Sea*. New York: Edward Elgar.

Hriestienko, Hank, and John E. McDonald. 2007. Going into the 21st Century: A Perspective on Trends and Controversies in the Management of the American Black Bear. *Ursus* 18 (1): 72–88.

Huffman, James L. 1995. In the Interests of Wildlife: Overcoming the Tradition of Public Rights. In *Wildlife in the Marketplace: The Political Economy Forum*, ed. Terry L. Anderson and Peter J. Hill. New York: Rowman & Littlefield.

Huggins, Laura E. 2013. *Environmental Entrepreneurship: Markets Meet the Environment in Unexpected Places*. Northampton, MA: Edward Elgar.

Ioannidis, John P.A. 2005. Why Most Published Research Findings Are False. *PLOS Medicine*. https://doi.org/10.1371/journal.pmed.0020124.

Joanen, Ted, Larry McNease, Ruth M. Elsey, and Mark Staton. 1997. The Commercial Consumptive Use of the American Alligator (Alligator Mississippiensis) in Louisiana. In *Harvesting Wild Species: Implications for Biodiversity Conservation*, ed. Curtis H. Freese. Baltimore: Johns Hopkins University Press.

Johnston, John, J.W. Ormsby, H.N. Campbell, Edward Silverman, H.T. Drake, Stephen Meunier, C.H. Mathews, Alfred James, and R.G. Richter. 1896. The Hunting of Deer. *Milwaukee Journal* (7 Nov.).

Jonker, Sandra A., Robert M. Muth, John F. Organ, Rodney R. Zwick, and William F. Siemer. 2006. Experiences with Beaver Damage and Attitudes of Massachusetts Residents Toward Beaver. *Wildlife Society Bulletin* 34 (4): 1009–1021.

Kessler, Winifred B. 2018. The Canadian Constitution and Wildlife Policy. In *North American Wildlife Policy and Law*, ed. Bruce D. Leopold, Winifred B. Kessler, and James L. Cummins. Boone and Crockett Club: Missoula.

Kreppel, Peter. 1897. About Duck Hunting. *Milwaukee Journal* (17 Feb.), 8.

Kuhn, Thomas. 1996. *The Structure of Scientific Revolutions*. 3rd ed. Chicago: Chicago University Press.

Laliberte, Andrea S., and William J. Ripple. 2004. Range Contractions of North American Carnivores and Ungulates. *BioScience* 54 (2): 123–138.

Laskow, Sarah. 2017. The Giant Frog Farms of the 1930s Were a Giant Failure. *Atlas Obscura* (25 Oct.)

Leal, Donald R. 1998. Cooperating on the Commons: Case Studies in Community Fisheries. In *Who Owns the Environment?* ed. Peter J. Hill and Roger E. Meiners. New York: Rowman & Littlefield.

Lebel, Francois, Christian Dussault, Ariane Masse, and Steve D. Cote. 2012. Influence of Habitat Features and Hunter Behavior on White-Tailed Deer Harvest. *Journal of Wildlife Management* 76 (7): 1431–1440.

Leopold, Bruce D., Winifred B. Kessler, and James L. Cummins. 2018. Preface. In *North American Wildlife Policy and Law*, ed. Bruce D. Leopold, Winifred B. Kessler, and James L. Cummins. Boone and Crockett Club: Missoula.

Lewiston Journal. 1892. Trapping Bears and Killing Them. *Bismarck Tribune* (2 Apr.).

Liu, Nengye, Cassandra Brooks, and Tianbao Qin. 2019. *Governing Marine Living Resources in the Polar Regions*. New York: Edward Elgar.

Ljung, Per E., Shawn J. Riley, Thomas A. Heberlein, and Go Ran Ericsson. 2012. Eat Prey and Love: Game-Meat Consumption and Attitudes Toward Hunting. *Wildlife Society Bulletin* 36 (4): 669–675.

Lueck, Dean L. 1995. The Economic Organization of Wildlife Institutions. In *Wildlife in the Marketplace: The Political Economy Forum*, ed. Terry L. Anderson and Peter J. Hill. New York: Rowman & Littlefield.

Malloy, Steven J. 2001. *Junk Science Judo: Self-Defense Against Health Scares & Scams*. Washington, DC: Cato Institute.

Meiners, Roger, Pierre Desrochers, and Andrew Morriss, eds. 2012. *Silent Spring at 50: The False Crises of Rachel Carson*. Washington, DC: CATO.

Mighels, Philip V. 1897. Hints for Young Trappers. *Salt Lake Semi-Weekly Tribune* (11 May), 13.

Miles, Nelson A. 1895. Hunting Large Game. *North American Review* 161 (467): 484–492.

New York Times. 1877. Hunting in Florida. *St. Louis Globe-Democrat* (14 Jan.), 9.

O. 1889. Deer Hunting. *Portland Oregonian* (13 Jan.).

Organ, John F. 2018. The North American Model of Wildlife Conservation. In *North American Wildlife Policy and Law*, ed. Bruce D. Leopold, Winifred B. Kessler, and James L. Cummins. Boone and Crockett Club: Missoula.

Organ, John F., et al. 2012. The North American Model of Wildlife Conservation. *The Wildlife Society Technical Review* 12-04.

Organ, John F., Thomas A. Decker, and Tanya M. Lama. 2016. The North American Model and Captive Cervid Facilities—What Is the Threat? *Wildlife Society Bulletin* 40 (1): 10–13.

Ostrom, Elinor. 1990. *Governing the Commons: The Evolution of Institutions for Collective Action*. New York: Cambridge University Press.

———. 1992. The Rudiments of a Theory of the Origins, Survival, and Performance of Common-Property Institutions. In *Making the Commons Work: Theory, Practice, and Policy*, ed. Daniel W. Bromley. San Francisco: Institute for Contemporary Studies.

Palmer, E. Laurence. 1939. Farm Forest Facts. *Cornell Rural School Leaflet* 33 (2): 1–32.

Patera, Pat. 1978. There's Big Money in the Secret Art of Frog Farming. *Mother Earth News*, July/Aug.

Pennington, Mark. 2011. *Robust Political Economy: Classical Liberalism and the Future of Public Policy*. Northampton, MA: Edward Elgar.

Peterson, Markus J. 1992. Whalers, Cetologists, Environmentalists, and the International Management of Whaling. *International Organization* 46 (1): 147–186.

Peterson, Markus J., M. Nils Peterson, and Tarla Rai Peterson. 2016. What Makes Wildlife Wild?: How Identity May Shape the Public Trust Versus Wildlife Privatization Debate. *Wildlife Society Bulletin* 40 (3): 428–435.

Philadelphia Press. 1883. An Old Frog Hunter. *St. Louis Globe-Democrat* (12 Dec.), 6.

Prukop, Joanna, and Ronald J. Regan. 2005. The Value of the North American Model of Wildlife Conservation: An International Association of Fish and Wildlife Agencies Position. *Wildlife Society Bulletin* 33 (1): 374–377.

Reid, Colin T., and Walters Nsoh. 2016. *The Privatisation of Biodiversity? New Approaches to Conservation Law*. New York: Edward Elgar.

Richards, John F. 2014. *The World Hunt: An Environmental History of the Commodification of Animals*. Berkeley: University of California Press.

Rodiek, Jon E. 1991. Introduction. In *Wildlife and Habitats in Managed Landscapes*, ed. Jon E. Rodiek and Eric G. Bolen. Washington, DC: Island Press.

Roth, Alvin. 2015. *Who Gets What and Why? The Hidden World of Matchmaking and Market Design*. New York: William Collins.

San Francisco Call. 1885. Traps and Trappers. *Galveston Daily News* (5 Nov.), 6.

San Francisco Examiner. 1890. Hunting for Mud Turtles. *Bismarck Tribune* (9 Apr.).

Savage, James. 1825. *The History of New England from 1630 to 1649 by John Winthrop*. Boston: Phelps and Farnham.

Schorr, Robert A., Paul M. Lukacs, and Justin A. Gude. 2014. The Montana Deer and Elk Hunting Population: The Importance of Cohort Group, License Price, and Population Demographics on Hunter Retention, Recruitment, and Population Change. *Journal of Wildlife Management* 78 (5): 944–952.

Schwabe, Kurt A., and Peter W. Schuhmann. 2002. Deer-Vehicle Collisions and Deer Value: An Analysis of Competing Literatures. *Wildlife Society Bulletin* 30 (2): 609–615.

Seay, Katharyn. 2019. Alligator Mississippiensis. *Animal Diversity Web*. https://animaldiversity.org/accounts/Alligator_mississippiensis/.

Shaw, Christopher W. 2019. *Money, Power, and the People: The American Struggle to Make Banking Democratic*. Chicago: University of Chicago Press.

Shields, G. O. 1887. Hunting the Grizzly. St. Louis Daily Globe-Democrat (31 Jul.), 28.

Shrader-Frechette, K.S. 1991. *Risk and Rationality: Philosophical Foundations for Populist Reforms*. Berkeley: University of California Press.

Simon, Julian. 1999. *Hoodwinking the Nation*. New Brunswick: Transaction Publishers.

Smith, Christian A. 2011. The Role of State Wildlife Professionals Under the Public Trust Doctrine. *Journal of Wildlife Management* 75 (7): 1539–1543.

Speck, Frank G. 1915. The Family Hunting Band as the Basis of Algonkian Social Organization. *American Anthropologist* 17 (2): 289–305.

St. Louis Globe-Democrat. 1895. Hunting Dakota Wolves. *Daily Inter Ocean* (2 June), 31.

Troyer, Dianna. 2019. Western Innovator: Frog 'Ranch' Keeps Owners Hopping. *Capital Press*, 12 July.

Vantassel, Stephen M., Tim L. Hiller, Kelly D.J. Powell, and Scott E. Hyngstrom. 2010. Using Advancements in Cable-Trapping to Overcome Barriers to Furbearer Management in the United States. *Journal of Wildlife Management* 74 (5): 934–939.

Vercauteren, Kurt C., Charles W. Anderson, Timothy R. Van Deelen, W. David Drake, David Walter, Stephen M. Vantassel, and Scott E. Hyngstrom. 2011. Regulated Commercial Harvest to Manage Overabundant White-Tailed Deer: An Idea to Consider? *Wildlife Society Bulletin* 35 (3): 185–194.

Wales, William. 1770. Journal of a Voyage, Made by Order of the Royal Society, to Churchill River, on the North-west Coast of Hudson's Bay. *Philosophical Transactions* 60: 100–136.

Warnock, Mary. 2015. *Critical Reflections on Ownership*. New York: Edward Elgar.

Waselkov, Gregory A. 1978. Evolution of Deer Hunting in the Eastern Woodlands. *Midcontinental Journal of Archaeology* 3 (1): 15–34.

Whitney, Leon F. 1931. The Raccoon and Its Hunting. *Journal of Mammalogy* 12 (1): 29–38.

Winkler, Richelle, and Keith Warnke. 2013. The Future of Hunting: An Age-Period-Cohort Analysis of Deer Hunter Decline. *Population and Environment* 34 (4): 460–480.

Wright, Charles. 1868a. Bears and Bear-Hunting. *The American Naturalist* 2 (3): 121–124.

———. 1868b. Deer and Deer-Hunting in Texas. *The American Naturalist* 2 (9): 466–476.

Wright, Robert E. 2010. *Bailouts: Public Money, Private Profit*. New York: Columbia University Press.

History of the North American Wildlife Conservation Model

Abstract Chapter 3: Reviews the history of wildlife management in North America, from American Indians through the development of the North American Wildlife Conservation Model in the late nineteenth century, and surveys the interplay between biology, ecology, markets, range, and population size of a dozen important types of wild animals. The key takeaway is that common pool problems, especially when combined with species-specific characteristics, like herding and flocking behaviors, that render human predation particularly inexpensive endanger wild game populations far more than the scientifically managed commercialization of their flesh does.

Keywords North American Wildlife Conservation Model • Scientific wildlife management • Game animals • Causes of species extinction • Causes of local species extirpation/range reduction • Common pool problem

As economist Thomas Sowell once wrote, "what new idea will seem plausible depends on what one already believes" (Sowell 2009, 7). According to NAWCM's origin story, North America's wild game populations were on the verge of collapse when NAWCM emerged and averted certain disaster. According to some accounts, deer, bear, and turkeys were almost

© The Author(s), under exclusive license to Springer Nature Switzerland AG 2022
R. E. Wright, *The History and Evolution of the North American Wildlife Conservation Model*,
https://doi.org/10.1007/978-3-031-06163-9_3

extirpated (Braverman 2015, 146). Accounts of game population decline are not without factual basis, most obviously in the case of bison and passenger pigeons, but for most species they appear to conflate local extirpation or range shrinkage with extinction. After the successful reintroduction of wapiti, all but four (bison, Carolina parakeets, jaguars, passenger pigeons) of the 282 identifiable species described in John Lawson's 1709 *A New Voyage to Carolina*, for example, can still be found in one or both of the Carolinas (Hairr 2011, 313, 320). Of the four no longer found in the Carolinas, only the two bird species went extinct, for the complex reasons described below. Unfortunately, a fifth species, the red wolf (*Canis rufus*) remains critically endangered despite efforts to re-invigorate it (https://www.nationalgeographic.com/animals/mammals/r/red-wolf/).

Reports of vanishing game were partly the product of those who wished to keep their honey holes secret (Anon. 1876b) and partly due to hunters being called liars when they reported plentiful game (Anon. 1897b). Not helping matters was the playful tradition of relating "tall tales" (Anon. 1895d), like the old hunter who claimed that ducks used to be so plentiful they had to stack three on top of each other to nest while wild rabbits (probably cottontails, *Sylvilagus floridanus*) used to enter houses and fall asleep in rocking chairs (Munkittrick 1893). The teller of another tall tale claimed that ducks were once so thick that he accidentally speared 37 of them when an improperly secured ramrod flew from his gun upon discharge (St. Paul Pioneer Press 1893). With little reliable statistical evidence of population, or even harvest, numbers to go by until well into the twentieth century (Waselkov 1978, 19; Zink 2014, 2), piecing together the stories of each major game species and area is difficult, but understanding what happened to key species and why, the aim of this section, will help to untangle the key cause of the NAWCM's success.

GAME MANAGEMENT BEFORE THE NAWCM

Although American Indians are sometimes implicated in the extinction of North America's megafauna (Baden et al. 1981; Koch and Barnosky 2006; Nagaoka et al. 2018), partly due to hunting techniques that killed hundreds of mega-ungulates within minutes (Wheat et al. 1972), they managed wild resources successfully enough to survive, even in the face of significant climatic shocks (Fiedel 2001), for millennia. How they managed wild game has remained largely a mystery, and in fact the question only arises if population densities are thought to exceed a given region's carrying capacity (Anderson 1998, 259). That usually evokes images of Cahokia or Mesoamerica, but in fact Indians inhabiting more northern climes, "hungry country" as it is often called, probably achieved saturation quickly. For example, parts of the Rockies, including what is today Yellowstone National Park, may have been overhunted when Lewis and Clark trekked through in 1804–1806 (Kay 1997; Laliberte and Ripple 2003; Brown 2018).

Some scholars believe that Indian game management was intimately linked with war (White 1983; Smalley 2016, 317–22). Tribes respected each other's hunting grounds when game was abundant, but when game populations fell in traditional hunting grounds, tribes pushed boundaries (Speck 1923, 458), often until they fomented military conflict with other tribes. War reduced both harvest rates and human demand, thereby allowing game populations to rebound (Swanson 2018, 27–28). When boundaries were respected, buffer zones between tribal hunting grounds allowed game populations to repopulate depleted areas (Laliberte and Ripple 2003; Foster and Cohen 2007, 36). Incursions onto tribal hunting lands by Euroamerican hunters and trappers Indians rebuffed as vigorously as possible on the grounds that they were trespassers, poachers, and treaty violators (Smalley 2016).

Some tribes that hunted deer communally (Pluckhahn et al. 2006, 269–71) successfully managed game populations through customs, like hunting or trapping an area only every other or every third year (Michelson 1921, 238–39; Kinietz 1940, 179; Speck and Eiseley 1942, 220; Smalley 2016, 307). Archaeological and historical evidence suggests that most deer hunting took place from late fall through early spring so that nursing fawns and their mothers were spared (Spiess et al. 2006, 162).

Some believe that Indian hunting taboos, especially negative or prohibitive ones (Young and Cutsforth 1928, 283), like that about killing only the quarry sought after, served to conserve game (Smith 1889; Richards 2014, 10–11). In Inuit folklore, powerful spirits readily punished those who flouted accepted hunting norms (Mancall 2013, 17–19). Such norms adapted to circumstances, too. For example, natives soon learned to continue using the harpoon and not guns when hunting marine mammals because their quarry often sank when shot (Reagan 1919–1921, 447).

The ancients believed that Artemis (Diana) or other gods would punish hunters who broke long-standing rules about which species could be hunted, when, and where (Sokos et al. 2014, 452–53). Unsurprisingly, modern Euroamerican hunters also held superstitious beliefs that limited harvest, like not hunting on Sunday or bringing more than 15 shells on a hunt (Young and Cutsforth 1928, 284). Superstitions about bad luck from poaching other men's traps and lands also imposed important constraints (Speck and Eiseley 1942, 231). Taboos may seem like weak enforcement mechanisms but the one against shooting white-tailed does was notoriously difficult to subdue (Kluender et al. 1992; Zink 2014, 7). Many Indian groups also seem to have relaxed prohibitions at times. Caddoans in eastern Texas usually harvested adult tom turkeys, but the bones of hens and jakes sometimes appear in the archaeological record (Perttula et al. 1982, 91–92). Some Indian groups held entire species in reserve for hard times. When Inuit near Nain harvested only 300 seals instead of the usual 2000, for example, they redoubled their efforts hunting caribou (Wheeler 1930, 455–56). In some areas, species like porcupine were left alone in good times so their numbers could rebound after being used as an emergency food source in bad times (Speck 1923, 458). In other places, only lost hunters were supposed to kill porcupines (Young and Cutsforth 1928, 284).

The most interesting Indian game management strategy, though, was the tradition in the northern woodlands of North America, likely even in the pre-Columbian period (Cooper 1939, 89; Speck and Eiseley 1939, 1942), of assigning relatively small family groups the exclusive right to manage specifically delineated and heritable tracts of fishing, hunting, and trapping land of several hundred square miles (Speck 1915, 1923; MacLeod 1922; Cooper 1929, 286). In other traditions, Indian hunters divided the tribe's territory into exclusive hunting tracts that individuals controlled for a season, year, or longer, with boundaries adjusted by elders when necessary (Cooper 1939, 75–78). Those and other ways of reducing the common pool problem worked until Euroamericans arrived and disregarded the Indians' rules and property claims (Baden et al. 1981; Huffman 1995, 32).

Interestingly, Indians fighting the imposition of state game laws in the late nineteenth and early twentieth centuries noted that they had as much interest in conservation as ranchers did in their herds (Speck 1915, 294). Despite often facing complex legal labyrinths and Bureau of Indian Affairs oversight, an increasing number of tribes now successfully control hunting and fishing access on tribal lands (Anderson 1998, 273–74; Brown 2016,

20; Blackwell 2018, 391), examples of an important phenomenon known as community-based natural resource management that has been successfully implemented in Namibia and elsewhere (Barnes et al. 2002).

The wild game management policies of the French, Spanish, Dutch, Swedes will not be explored in detail here. Of the European nations to colonize North America, only Britain managed to achieve colonist numbers sufficient to impact wild game populations directly. Moreover, while the influence of other colonizers may still exert some influence in jurisdictions like California, Florida, Louisiana, Quebec, Texas, and of course especially Mexico and Central America (Gallardo 2018, 353), for the most part the NAWCM is of Anglo-American origin and strongest in Canada and the United States (Organ et al. 2012, 3; Leopold 2018, 25; Organ 2018, 126).

In Britain, game management was a function of state and often entailed the complete prohibition of hunting by all except the rich and powerful, especially on extensive state game preserves like the Wirral, across the Mersey from Liverpool. Even the collection of sticks by commoners was a punishable offense and poaching a beast of the forest or of the chase (big game animals) was punished by mutilation and even death (Lueck 1995, 3; Smalley 2016, 308–9; Leopold 2018, 19–23).

In North America, though, the British Law of the Forest transformed into the natural right of any person to kill "any wild creature" s/he could (Savage 1933, 32–33). Unlike in England, in America animals were considered "free for the taking" (Smalley 2016, 309–10). Before the US Civil War, Americans generally did not go into the field or stream for sport, like the English, but rather sought meat and other animal products. "Sporting is with us," explained the North American reviewer of numerous English books on fishing and hunting, "for the most part, not an art but a trade, and needs no teacher but personal experience" (Anon. 1842, 345). Because they hunted for the cooking pot rather than the wall, early American hunters often shot whatever wildlife they encountered on the supposition that it was either edible or that it preyed on edible animals. Subsistence hunting alone, however, could not extirpate wild game because as prey became less abundant, it essentially became more expensive as it took longer to find and kill, inducing people at the margin to buy farmed fare instead (Berthel 1935, 262). Commercial hunters could be driven to hunt for scarce game if market prices moved high enough, but most consumers switched to domesticated animals instead (Braverman

2015, 147). Commercial hunting for mounting specimens, as for owls, was unlikely to lead to extinction because demand was necessarily limited as mounts lasted for years and relatively few people wanted them (New York Mail 1887).

Colonial and early state governments passed hunting laws, but they were neither stringent nor well enforced, especially where game remained abundant (Berthel 1935, 264; Leopold 2018, 25). The notion that all wild animals were fair game translated into hunting techniques that were more about efficiency than sport. Moreover, many North Americans cared more about economic growth and national expansion than the fate of other species, or even other human groups. It was common for Americans to assert that the peopling of the Great Plains and the West more generally would lead to the extinction of Indians as well as most animal species. As R. B. Marcy (Anon. 1866a, 581) put it, "the wild animals that abound on the great plains to-day will soon be as unknown as the Indian hunters who have for centuries pursued them." The decline of the great bison herds, even when lamented, was looked upon with favor because their extermination made room for cattle (Dunraven 1879, 361) and kept Indians on their reservations (Anon. 1886d). Nelson Miles put it bluntly after the carnage was done: "The buffalo [sic], like the Indian, stood in the way of civilization and the path of progress, and the decree had gone forth that they must give way" (Miles 1895, 492). The expectation and acceptance of eradication allowed unregulated commercial harvest to persist.

As noted above, one powerful wildlife management technique used by Indians entailed restricting access to hunting and trapping land. Until the late nineteenth century, by contrast, American hunters could enter unenclosed private land *without landowner permission* throughout much of the nation (Biber and Eagle 2015, 820 n. 186). Colonists considered wild game a common good as there was no alternative, "it not being possible to make Inclosures in the Woods" given the high cost of labor and the then state of technology (Millan 1744, 25).[1]

In open-range states, landowners had to enclose their land to prevent grazers from entering it. The owners of grazers did not have to fence their herds in, so few took the expense. That allowed hunters to roam lawfully over large tracts of government and private land in search of quarry. And roam they did, even when farmers posted no trespassing signs (Anon.

---

[1] Indeed, fence building was a major component of physical capital formation in the United States even in the nineteenth century (Gallman and Rhode 2019).

1891a). Importantly, however, the large commons, even combined with the commercialization of most wild game species, continued for centuries before negatively impacting total populations.

THE EMERGENCE OF THE NAWCM

By the late nineteenth century, the tenets of the NAWCM were emerging. Sportsmen, alone and in clubs like Boone and Crockett (Leopold et al. 2018), deplored the dearth of game and feared a repeat of what happened in the West with the rapid destruction of bison as well as mountain ptarmigan (*Lagopus leucura*) and heath hen (*Tympanuchus cupido*) (Ferril 1893). More game laws were passed, though not always terribly scientific ones, at least at first. One Wyoming trapper complained that the ban on marten trapping had been passed by urbanites and plains ranchers who thought they had banned the trapping of martin, probably *Progne subis*, but definitely a type of small migratory bird in the family *Hirundinidae* (Thomas 1906, 50). Such laws hardly inspired confidence in the emerging science of wildlife management. Many deer hunters, for example, wondered why they could not hunt with dogs during a season of limited duration with a two-tag limit (Moore 1897).

Most states started by banning the most outrageous methods of harvest, like poisoning fish or using explosives to kill or stun and then net them (Anon. 1885a). Others, like Missouri, enacted closed seasons enforced by monitoring markets and with help from private associations like the Missouri Fish and Game Protective Association (Anon. 1881e).

Instead of a few market hunters harvesting large numbers of animals for sale in major markets, urbanites began to take to the field to bag and tag their own game, often for the meat but increasingly for the sport and comradery. While commercial hunters harvested far more animals than sport hunters on average, the sportsmen far outnumbered the commercial hunters. So it remains unclear that banning wild meat markets would have had much of an effect on harvest rates had not a sportsman's ethos evolved in the late nineteenth century.

As one Indian complained, a white hunter will kill "everything he can find, whether he needs its flesh or not, and then when all the animals in one section are killed he takes the train and goes to another where he can do the same" (as quoted in Speck 1915, 294). Indeed, an emerging group of wealthy sport hunters actively sought out more fertile hunting grounds in the backwoods of their own states (Williams 1889) as well as further afield, including northern Maine and Minnesota (Anon. 1876b), the swamps of Florida (New York Times 1877; Reagan 1919–1921, 444), the more remote parts of the Rockies (Chicago Times 1888), and in Alaska, Africa, Asia, Mexico, and Canada (Anon. 1877, 49; 1895a, 1897a, 1923; Burroughs 1915, 139).

Contemporaries noted that Britons and Americans both seemed "to regard it as the whole duty of man to go somewhere and kill something" (Williams 1889). Indeed, one group of nineteenth-century hunters bragged of killing or wounding all but a brace of an entire covey of quail (New York Times 1877). Others killed "as many moose and deer as they desired" (Anon. 1891c) and a pair of hunters in Colorado once harvested 18 wapiti in 10 minutes (Fort Collins Express 1881). A fellow in Milwaukee shot ducks off the city pier, mowing them "down in a bunch" whenever the fancy struck or his pot was empty (Anon. 1896a). Another water-fowler aimed to set the record for killing the most canvasback ducks, part of an advertising campaign to induce "gentlemen from the North" to hunt near Galveston, Texas (Anon. 1896b).

Late nineteenth-century hunting stories almost always stressed quantity over quality (Anon. 1895f). The leaders of rival mining camps, for example, bragged about who killed more deer for camp meat (Anon. 1876c), while a debate raged over whether South Carolina's General Wade Hampton or one-armed California shepherd H. C. Hanson was America's mightiest hunter. The former claimed to have killed 500 black bears and 16 mountain lions and the latter 512 bears, mostly browns, plus 308 mountain lions "besides other game" (Anon. 1894a). The mightiest hunter around Denver was said to be J. H. Marden because he regularly hauled a wagon load of deer, mountain lions, and wapiti out of the moun-tains (Denver News 1883). Jake Hammersley was legendary in Pennsylvania for killing over 2000 "deer, bear, panther, and smaller game" with his single-shot muzzleloader (Erie Dispatch 1884). Jimmy Todd in the Valley of Virginia had a similar story, taking "over 2700 deer" and "bears with-out number … with one old muzzle-loading rifle" (Staunton Vindicator 1877). Born in Maine but later a resident of upstate New York, John Hutchins boasted of killing 100 moose (*Alces alces*), 1000 deer, 10 cari-bou, 100 bears, 50 wolves, 500 foxes, 100 raccoons, 25 wild cats, 100 lynx, 150 otter (*Lutrinae* sp.), 500 beaver, and 400 fishers (*Martes pen-nanti*), other assorted furbearers by the thousands, and muskrats "by the ten thousands." He claimed that hunting and trapping earned him between $5 and $75 a month (Trapper's Guide 1865). Not to be outdone, D. P. Graves measured his kills by *the ton*, claiming to harvest three tons of deer in a single season in Wisconsin (Anon. 1886e).

Yet another mighty hunter bragged of killing 120 wapiti and 200 deer near Mt. Hood and claimed to net $2000 to $3000 per year at a time when a farm laborer made $225 a year, without board (Anon. 1893a). Nathan B. Moore made the papers because he killed 275 moose by age 68 (Lewiston Journal 1886). In New York's Adirondack Mountains, Warren Hume killed over 4000 deer and 500 bears during his career, and once killed 7 deer in a single day. He made between $500 and $800 a year (West Bloomfield 1889).

When North Americans expected animals would go extinct or be driven from inhabited areas, they did not quibble about the methods used to take game. In the late nineteenth century, men who considered themselves sportsmen would take deer at night, even in the middle of summer (Campbell 1889a), with crude lights fashioned from old frying pans and pine resin (Audobon's Ornithological Biography 1838; Wright 1868b, 472–73; Special Correspondence 1887), on land or by boat. Sitting over a deer "lick" was considered high sport (Erie Dispatch 1884). Train engineers thought nothing of running down deer caught on their tracks (Anon. 1886a, 354) and hunters speared entire families of muskrats in their dens (Clothier and Furnisher 1890; Baltimore Sun 1892). They also flooded them out of their dens or shot them traveling in the moonlight (Anon. 1885b). Moose, including cows, were taken by moonlight (Dunraven 1879, 347, 353, 365–66; Dunraven 1881; New York Times 1884) and torch light (Oxley 1888; Williams 1889) or by trapping them in deep snow in their "yards" in the dead of winter (New York Times 1884; London Saturday Review 1887), and tracking them for days on end if they managed to escape the yard (Anon. 1895e). Caribou and deer were killed with knives by "crusting" them, that is, driving them into deep drifts grown soft in the spring sun (Special Correspondence 1887; Williams 1889).

Alligator numbers in Florida took a hit when night hunting for them was conducted by tourists and suppliers of the alligator skin and teeth markets (New York Times 1877; Anon. 1881a, 279). Turkeys were also hunted at night (Ferril 1893) and the fairest way of shooting a gobbler in the nineteenth century, it was said, was on the roost (Anon. 1886b)! Others used tame turkeys to decoy and call wild ones (St. Louis Globe-Democrat 1894).

Night hunting methods often culled females and the young instead of old males. One observer warned that instead of shooting deer, night hunters sometimes shot at stars or other hunters (Anon. 1890) and too often killed "cows, horses and other domestic animals" (Audobon's Ornithological Biography 1838; Campbell 1889b).

Many hunters, even in the north, used dogs to run down deer and moose (Lewiston Journal 1886). Carley (1897) attributed declining deer populations in New York and the upper Midwest to the use of dogs. On the high plains, some hunters tried to spot and stalk pronghorn, missing many long shots in the process, but many others used greyhounds to run "antelope" down and even kill them (Batty 1874, 38; Roosevelt 1892). Using packs of dogs not just to find but fight and kill wolves, bears, and coons was considered a "royal sport" (Miles 1895, 491; Anon. 1899a) in which even women partook (Anon. 1899c). Hunters killed wapiti "with a thoughtless brutality" that threatened their "extermination in civilized districts" (Dunraven 1879, 339). One technique, used by Californios, entailed running down herds with horses, lassoing and then "talking sarcastically to the struggling elk" before dispatching it with a long, curved knife at the end of a spear called a luna (Scanland 1893).

When modern rifles, smokeless powders, and ballistic bullets began to come out around the turn of the century, extending effective ranges from the traditional 200 yards to over a thousand, hunters were more worried about shooting each other than the ethics of harvesting animals at such great distances (New York Sun 1897; Anon. 1898a, 85).

Restricting harvest methods alone, if it could have been enforced, might have been enough to slow or even reverse game population decline. Effectively restricting methods raises the cost of harvest, which ceteris paribus would have led to a reduced quantity being brought to market.

Nevertheless, the methods used to harvest game is not so important as the number harvested. A pronghorn killed by greyhounds is just as dead as one killed along an "antelope fence" constructed by Indians in Utah or New Mexico (Anon. 1895g; Reagan 1919–1921, 443). Many of the night-hunting techniques, like the use of torches to attract or catch the eye shine of curious moose and deer, were regularly used by Indians (Reagan 1919–1921, 444), who, like Euroamericans, also killed moose made immobile by deep snow (Oxley 1888). As discussed above, however, Indians tended to limit harvest numbers by other means, killing what they needed as efficiently as possible (Barboza and Tihanyi 2018, 380). Some Euroamerican hunters, by contrast, killed as many bison as they could, just to gloat over their prowess (Western Hunter 1889; Branch 1929). While it is a myth that Indians always used every part of every bison (or other animals, cf. Richards 2014, 45)—they often just ate the tongues or the unborn—they typically limited the number they harvested, and that made all the difference.

While it would be easy to call for more ethical hunting practices, ethics is a tricky, normative issue. One technique for capturing wild mustangs, shooting them just under the spine, knocking them out long enough to hobble them, does not seem ethical today. (Many recovered but some, shot too high, did not [Kansas City Star 1890]). But waterfowling techniques that today are both legal and considered sporting were once condemned by purists who argued that the use of decoys was "shameful" (Harper's Magazine 1870) and that any man who would decoy ducks was also capable of stealing sheep (Forest and Stream 1882). Even camouflage clothing was long suspect (Anon. 1918, 383).

Eventually, sport hunters disappointed on too many hunts grew weary of the slaughter unleashed by pot hunters and moved to ban the most effective hunting techniques and commercial hunting (Kreppel 1897). Even after wild meat markets had been banned, though, markets remained a crucial part of hunting. The markets changed, though, and increasingly aimed at luring hunters and other consumptive users to specific states or towns. Advertisements began to appear, touting destinations like Maine, where deer, fish, and moose always abounded, at least according to the railroads that in the age before automobiles and paved roads were the main means of passenger travel (Barker 1896; Anon. 1898b, 1899b). Hunters numbered in the hundreds, and the daily return of deer ran in the scores, with moose being fewer but "grand specimens" (Whig Man 1899; Anon. 1899b, 1901, 246; 1902, 291).

As the price of protein plummeted and urbanization continued apace, hunting for food decreased, while hunting for sport increased. Already by the 1920s, observers believed "that game laws, and commercial foods insure good hunting in remote places to future generations" (Eddy 1924). Books about hunting "trophies" using "ethical" hunting tactics soon began to appear and old market hunters, trappers, and Indian scouts increasingly turned to guiding (Thomas 1906, 117–18; Rockwell 1922, 12), a practice that began in the nineteenth century when sport hunters began to solicit advice from commercial hunters, some of whom established private hunting regulations, essentially trading hunting information for compliance with rules they believed would maintain a sustainable harvest in the areas they regularly hunted (F.E.S. 1878; Dunraven 1881; New Orleans Times Democrat 1883). Even in the nineteenth century, most guides followed game laws because they realized that their livelihoods depended not so much on their awkward, unskilled clients, mostly "city sporting gentlemen," killing game as them *seeing* animals (Special Correspondence 1887). In fact, it was the guides who de facto ended "hounding" in the Adirondacks in the 1880s (Special Correspondence 1887). Importantly, commentators began to deride "the usual narrative of animals slaughtered or left to die" (Burroughs 1915, 139) and the use of dogs and other methods of chase slowly came to be seen as unfair, unethical, and counterproductive in many areas (Carley 1897). When successful hunts only filled hunters' hearts with pride, instead of their bellies with food, harvesting animals by any means necessary no longer made sense (Hunt with the Yankton Sioux 1873). But it took time for attitudes to change. When they did, sportsmen used their superior numbers to pass laws discouraging and then banning commercial hunting (Baltimore American 1887).

Rising interest in hunting brown bears (*Ursus arctos* formerly *Ursus horribilis*), long considered dangerous and expensive to hunt and good only for their hides, helps to identify the shifting ethos. Shields (1887), for example, argued that summer hunting and trapping of brown bears, which was sometimes done by affixing string to bait and a gun trigger, was unsporting. "Game of any kind," he added, "should always be pursued in a fair, manly manner." Others mocked "brave hunters" who purchased captive bears and shot them while still chained (Forest and Stream 1889). By 1893, James Newton Basket openly declared that shooting young squirrels in May was "scarcely a sportsmanlike proceeding and by no means has a tendency to the game's preservation."

Mass carnage remained the name of the game in some circles though, and stories of sows and all their cubs, along with a stray bald eagle and an unrecoverable drake harlequin (*Histrionicus histrionicus*), being massacred by hunting parties were still told with pride in the early twentieth century. By the 1920s, brown bears were scarce enough even in remote parts of Alaska that the governor could credibly call for their extermination even while insurers refused to issue brown bear hunt coverage.

Compounding matters was the desire of naturalists to obtain a specimen before it was "too late" (Rockwell 1922). One such, for example, slaughtered numerous mountain sheep (*Ovis dalli*) because a museum wanted to display an entire family, from lamb to ewe to juvenile to full curve ram. In the process, he killed his eighth brown bear, a trophy two-year old reminiscent of the wapiti bull calves that Theodore Roosevelt once bragged of bagging (Anon. 1886d).

By the final quarter of the nineteenth century, wealthy Americans in the northeast hunted foxes and imported stags with hounds for sport, on horseback no less, like proper English lords and ladies (New York Commercial 1875; Anon. 1894b, 1913, 51–52; Crossways 1896; New York Tribune 1896; Savage 1933, 30). The whole scene was already being lampooned in the West by the late 1880s (Nye 1887). Southerners also enjoyed the sport (A Hunter 1881; Anon. 1886c), which was thought to be good exercise for middle-aged men. Similarly, hunting was touted as a form of education for teen boys who increasingly hunted more for adventure than out of hunger (Curtis 1914, 260–61).

Poaching undoubtedly occurred (Ferril 1893). Sea otters, for example, were thought to be "on the verge of extinction and unless drastic measures are taken to enforce the present laws, it will be next on the list of wild creatures that we regret exist no more" (Rockwell 1923, 73). Most people, though, began to follow the new games laws and hunted where their target species remained abundant. Or, they adapted and began to "hunt" with cameras (Nesbit 1926; Berthel 1935, 259), counting pictures as hard-won "trophies" (Forest and Stream 1892; Kellogg 1926, 116).

Over time, compliance increased further as wildlife managers proved that they could increase wild game populations through careful management of variables like bag limits and estimates of unbagged mortality, cull gender, habitat, season length, and stocking size, especially where detailed data from hunter surveys could be collected (e.g., Baumgartner 1942; Burroughs and Dayton 1941, 159).

Natural Histories of Some Game Animals

The next part of this section argues that many game animals (beaver and other furbearers, deer, bear, large ungulates, small game, turkey, and waterfowl) often thought to have been brought to near extinction were never seriously endangered even though their flesh and other body parts were sold commercially. Unfortunately, pre-World War II population estimates are not available for most species, including white-tailed deer (Barboza and Tihanyi 2018, 377), in most places (Brook and Bradshaw 2006). A common proxy, harvest totals, are spotty and mostly postwar-only as well (Kluender et al. 1992). Moreover, even harvest totals drawn from private hunting estates in Europe inaccurately measure game populations unless hunting effort (total hours hunted) is also accounted for, data completely lacking for the prewar United States (Imperio et al. 2010).

The evidence adduced here, therefore, must remain anecdotal rather than conclusive, but it is an improvement on earlier efforts like the oft-cited Trefethen (1975). Moreover, the historical method of scanning newspapers and books for pertinent observations employed herein is not very different from that employed in studies like Webb (2018), which search the internet for what is more properly termed capta (the thing taken) than data (the thing given). In addition, assuming that the conclusions of modern studies (e.g., Fryxell et al. 1991) can apply to conditions just a century ago suggests that range reduction (local extirpation) was the main problem and that it was caused as much by habitat loss as by commercial hunting.

Even today, population trends are difficult to discern from natural fluctuations due to changes in habitat and interactions with non-human predators, but the available evidence does suggest a population V, bottoming out circa 1900, for many game species, including white-tailed deer (see, e.g., Webb 2018). At issue here is the depth of the V at its lowest point. The balance of the evidence suggests that it was not as low as some proponents of wild game meat bans assume based on much less evidence than presented here. One recent study, for example, claims that "the market hunter left many once-abundant species teetering on the brink of extinction." That same study also asserts that "unregulated trafficking ... in the 19th century led to drastic and, in some instances, catastrophic declines in populations" even though unregulated commercial harvest had been conducted continuously for *centuries* without causing large wild game population declines (Organ et al. 2012, 3, 14–15).

That of course points to the issue of causation. Was the V-shape population pattern due mainly to commercial harvest or to a complex confluence of factors? The balance of the evidence discovered and presented below suggests that those species that went extinct, or nearly so, were killed off not due to markets alone, but by habitat shrinkage or degradation and incentives to harvest combined with the low cost of harvest due to improved technologies, exacerbated by herding (bison) or flocking (carrier pigeons) behaviors. One extinct bird (Carolina parakeets) may have succumbed to a pathogen and waterfowl, especially divers and some dabblers (Callaghan et al. 1997, 519–27), suffered mostly due to the use of lead shot and nets capable of capturing entire flocks. Whales, by contrast, were endangered by a combination of commercialization, technological advancement, and an international tragedy of the commons. Table 3.1 summarizes the narratives that follow.

Table 3.1 Summary of the natural history of major North American game animals

Species or group	Commercialization	Outcome	Current status
Beaver	Pelts and meat	Local extirpations but never endangered	Thriving and local nuisance
Other furbearers	Pelts and meat	Range changes but most never endangered	Generally thriving
Small game	Meat	Range changes but most never endangered	Generally thriving
Whales	Meat, blubber, oils, other products	Near extinction due to international tragedy of the commons	Rebounding due to bans and pressure from environmental groups
White-tailed deer	Hides, meat, antlers	Greatly reduced numbers due to habitat destruction and tragedy of the commons	Pre-Columbian population levels achieved; widespread instances of local overabundance
Turkeys	Meat	Reduced range and numbers due to habitat destruction and common pool problem	Thriving over re-expanded range; local nuisance in some areas
Bears	Hides and meat (primarily black bear)	Range reduction	Generally thriving and some local nuisance issues
Large ungulates	Meat and hides	Range reduction due to habitat destruction	Elk, moose, caribou steady to thriving but possibly some issues due to global climate change
Bison	Hides and meat	Near extinction caused by common pool problem and herding behavior	Thriving public and private herds but far from historical population levels
Passenger pigeon	Entire bird	Extinction caused by common pool problem and flocking behavior	Close substitutes and difficulty breeding in captivity meant they were never domesticated
Carolina parakeet	Feathers	Extinction possibly precipitated by a pathogen	Unexpectedly rapid decline meant domestication efforts came too late

Sources: Narratives presented below

Castor Canadensis *and Other Furbearers*

Contrary to common belief (Brown 2018, 5), furbearers were not almost exterminated by trappers in the antebellum era. A negative demand shock caused a decrease in price that led to trapper exit. Beaver harvests dropped in the late 1830s, for example, due to global recession, an exogenous change in consumer preferences (Richards 2014, 5–8), and input substitution in the form of silk and rabbit fur (Anon. 1866d, 1870a). Although there is widespread belief that beavers, which numbered perhaps 100 million before European contact, were almost exterminated (Jonker et al. 2006, 1009), that surely was not the case. Figuring a harvest of 250 beavers per year per man (Anon. 1888), trappers, Euroamerican and Indian combined, were simply too few, given the state of technology and infrastructure in the early nineteenth-century North American west, to extirpate beavers on anything larger than a local level (Anon. 1866d; Richards 2014, 11, 32, 47–50; Wright 2019). Moreover, where the beaver trade was monopolized, as in parts of Canada, beaver harvests were held just below maximum sustainable yield by market incentives (Carlos and Lewis 1995, 83).

Total annual beaver harvests dropped from the hundreds of thousands in the eighteenth century to an average above 75,000 in the nineteenth as new areas were exploited and older areas were repopulated from adjacent and remnant populations (Richards 2014, 50). By the 1850s, trappable populations haunted Louisiana's bayous (Natchez Free Trader 1852). In the 1870s, beavers were so numerous in Colorado that big game hunters had to break up their dams in order to float hunt down dammed smaller rivers (Dunraven 1879, 344). That same decade, trappers made a good living trapping beavers in the Florida panhandle, Georgia, and Alabama. In addition to cash for the furs, they received ten cents a pound for the meat in the market in Rome, Georgia (Gainesville Florida Eagle 1880). By the 1880s, a colony of beavers large enough to produce 30 furs was trapped within 30 miles of St. Louis, Missouri (Anon. 1883b). In the 1890s, beaver trapping remained profitable in Walla Walla County Washington and elsewhere (Anon. 1893b; Cathlamet Gazette 1891). Trappers appreciated beaver because they supplied food for man and dog and their traps rarely attracted less desirable species (San Francisco Call 1885).

Most other furbearers were harvested in larger numbers in the nineteenth century than in the eighteenth. Only bear, otter, and wolf suffered large declines in the nineteenth century (Richards 2014, 49). Although trappers can certainly check furbearer populations, habitat loss has played a bigger role in the range reduction of iconic species like marten (Bissonette et al. 1991, 117–23).

In the north country, the Hudson Bay Company continued operations as usual, trading pelts caught by Indians for sundry manufactured goods (Robinson 1879). Indeed, Indians throughout the continent continued to barter and sell furs for manufactured goods (Anon. 1870b). As late as the Great War, Indian trappers in Minnesota annually deadfall trapped and snared animals with furs worth thousands of dollars (Reagan 1919–1921, 444–45).

Trappers shifted areas and species as prices and populations dictated (Anon. 1884a; San Francisco Call 1885), supplementing their income in the off-season with guiding and packing (Forest and Stream 1895). In the 1840s, George W. Pitts made a good living in Indiana trapping raccoons, fox, and otter by baiting his traps with leftover parts of the squirrels (various *Sciuridae*) and turkeys he consumed and concentrating on less accessible areas along rivers (Indianapolis Journal 1886). In 1861, Dr. Burnham and a helper made large sums trapping high points in the Sierras for fox and other furbearers in fashion (Mountain Messenger). About that same time, marten trappers kept busy keeping wealthy women warm and in fashion, many by using inexpensive deadfall traps produced on site (Anon. 1866b, 1866c). Trappers targeting other species also produced various snares and pit traps from available materials to keep their overhead low (Chicago Record 1896). Books about how to trap and care for pelts helped greenhorns begin to learn the craft (Williams and Bugbee 1865; Anon. 1867) and the Oneida Community near Utica, New York, thrived after the Civil War by supplying farm boys throughout North America with effective, inexpensive traps (Hodak and Masterson 2021).

In 1874, one observer explicitly stated that because fur prices were three times higher than in recent years, "many who had quit the business for more remunerative employment, have again started." Moreover, "the high prices seem to be caused more by an increase of the demand than a falling off of the supply, for the rivers, creeks, and mountains of Montana abound with game valuable for pelts" (Anon.).

In the 1880s, people in rural Connecticut could make more money trapping skunks or "polecats" (probably slang for *Mephitis mephitis* but possibly *Mustela putorius*) than through any other activity. Deadfalls were set near dwellings and outbuildings and bait consisted of chicken wings and other less desirable animal parts, so input costs were low. The animals were also hunted with clubs. In addition to their furs, which found a ready market of around a dollar (up from 15 cents 20 years earlier [Anon. 1873b]) depending on size, color, and quality, their oils were believed to have medicinal qualities (Holmes 1893, 218). During the Great Depression, New Yorkers harvested 42,699 skunks (Palmer 1939, 17).

Trapping continued in Michigan too, but the relative scarcity of game induced many trappers to exit because the incomes of the least efficient trappers dropped too low when fur prices did not keep pace with the amount of investment, monetary and temporal, needed to catch furbearers. Some supplemented their incomes, though, by collecting ginseng or honey (Chicago Record 1896). Trapping in California became relatively less important than formerly, but in the 1870s and 1880s "animals of the fine fur" were said to be "just as plentiful" as previously (Chico Chronicle 1886).

The muskrat trade in Wisconsin was brisk in the early postbellum era, "assuming an importance not generally known." Amos Benedict bought up 2000 acres of marsh for a nominal sum in order to quasi-ranch them, producing about 3500 skins per season. Other trappers caught just as many on public lands about the Fond du Lac and Winneconne Lake (Winnebago County Press 1870). In the 1890s, when muskrat pelts went for about a quarter a piece, "thousands of bales of rat skins are collected and shipped from all the Northern states" (Anon. 1895h), and some Marylanders (Clothier and Furnisher 1890) and New Jerseyites (Anon. 1885b) made a modest living by trapping muskrats.

In the 1870s, observers noted that Lewiston, Maine, had emerged as a fur trading center (Anon. 1873b). Although good furbearing range shrank over time in Maine (New York Times 1884), Maine's fur trade was worth $100,000 a year in the 1880s (Lewiston Journal 1881). Trapping remained viable in Maine in the 1890s "despite the hordes of hunters who flock into the state from all over the Union." Each spring, dozens of trappers made a "snug sum" bringing to market the pelts of bear, beavers, martens, and other furbearers (Anon. 1896c). Furs totaling over $150,000 were brought in some years in the 1890s. A drought made muskrat and fox easy targets for Maine trappers in 1899 (Anon. 1899).

In the 1890s, trapping in North Carolina could be so lucrative that four men gave up cotton farming to form a trapping company that hauled in hundreds of otters, minks (*Neovison vison*), weasels (*Mustela* sp.), and other furbearers (Charlotte News 1894).

Raccoon trappers made enough money during Louisiana's short winter to hold them over the entire year as they were able to get from 15 to 50 cents per skin, plus 5 to 10 cents for each skinned carcass from hungry African-Americans (Anon. 1896d). Throughout the nineteenth and well into the twentieth century, people throughout the continental United States trapped and hunted raccoons in order to sell their furs for the manufacture of clothing, especially coats (Anon. 1899a). Raccoons from the South with lighter furs composed the body of such coats, while coons with heavier coats from the North were used for trim (Whitney 1931, 30). Typically, hunters employed dogs and one who counted himself among the best took 128 over five years, including 45 in his best season, while a group of 3 hunters, including that same man, harvested 300 raccoons over 7 seasons (Whitney 1931, 29, 31). This hunter believed that strong natural selection in favor of intelligence was at work, leading to the survival of many coons caught in traps and hunted by the best dogs (Whitney 1931, 36–37).

Trapping remained difficult work so strong incentives to ranch furbearers existed. Once furbearers could be successfully ranched, fur farms soon flooded markets with mink, black raccoon (Whitney 1931, 31–32), and silver fox. By the 1920s (Anon. 1925a, 306), fur farming was so important in Alaska that many Alaskans shot bald eagles (*Haliaeetus leucocephalus*) on sight because they raided fur farms from above (Rockwell 1923, 71).[2] While trapping continues to this day, it is a shell of its former self because of fur farms, not because of the desolation of furbearing species.

Cetaceans

Commercial whaling also continued after its mid-nineteenth century heyday, well into the twentieth century in fact (Murdoch 1917). Fuel oil came from whale blubber and superior candle wax from spermaceti. Whaling declined in importance in America after 1860 partly due to the shocks the industry suffered during the Civil War (Mulderink 2012, 138–63), but mostly because substitute fuels and waxes were discovered and became much cheaper than sailing the seven seas for enormous marine mammals (Ross 1979; Archibald 2013). Americans never developed a taste for whale meat and by the postbellum period had no need for it (Shoemaker 2005).

---

[2] Hunters and farmers long considered eagles a scourge for stealing wounded game and young hogs (Hairr 2011, 318).

Still harvested for ambergris, blubber, and baleen (Reeves and Smith 2006), whales remained subject to an international tragedy of the commons exacerbated by technological improvements in the form of steam ships, cannon harpoons, and ramped factory ships. Attempts at international regulation in the 1930s in the form of sanctuaries, outright bans on taking Right (*Eubalaena sp.*) and Gray (*Eschrichtius robustus*) whales, and size and sex limits on other species, largely failed. After World War II, a new international treaty and regulator, the International Whaling Commission (IWC), exacerbated the tragedy of the commons problem among signatories (McHugh 1977) and cheating by the Soviet Union worsened matters further (Walsh 1999). By the 1970s, many whale species were so depleted (approximately 99 percent according to Clapham 2016) that the industry collapsed, allowing environmentalists from non-whaling countries to take control of the IWC.

Since 1986, the IWC, over the protests of Japan and Norway, has banned commercial whaling. Exclusions for scientific and traditional subsistence whaling, the resumption of commercial whaling by Norway in 1993, and continued whaling by Iceland and Russia, however, mean that each year thousands of large whales, and tens of thousands of smaller cetaceans not subject to IWC jurisdiction, are subject to human predation while their populations also come under stress due to climate change and pollution (Ottaway 2013). The United States banned commercial whaling in 1970 and is a staunch supporter of the IWC's ban, perhaps because of the perceived effectiveness of the wild meat market ban in the NAWCM (Gillespie 2005; Carlarne 2005). Catch shares, like those used to help fisheries, could help manage whales and could even be transferable to anti-whaling groups, "but more consensus on the matter is needed" (Anderson and Libecap 2014, 195–201).

Odocoileus Virginianus

White-tailed deer, the hides of which American Indians supplied to Euroamerican colonists in the hundreds of thousands annually throughout the eighteenth century, survived commercialization for centuries (Pressly 2013, 194–95; Richards 2014, 33–40). For millennia, Indians had hunted deer (and other cervid species, including moose, mule deer, and wapiti, where extant) via various techniques for their own consumption (Waselkov 1978; Perttula et al. 1982; Madrigal and Holt 2002; Spiess et al. 2006; Diamond et al. 2016) but deer remained numerous, up to 30 million continent-wide (Smalley 2016, 306; Swanson 2018, 13).

Indians increased per capita production in order to obtain European-manufactured goods at approximately the same time the number of Indians declined due to smallpox epidemics, so the annual harvest likely did not change dramatically (Zink 2014, 2–5). Indians also supplied Euroamerican markets with clay, ginseng, slaves, and other goods, so the full brunt of Indian balance of payment deficits did not fall on deer (Dunaway 1994). Nor did deer have to compete with hogs and other domesticates for food in Indian territory because most tribes relied on venison and other wild game for protein until the early nineteenth century (Pavao-Zuckerman 2007).

Economist historians (Mancall and Weiss 1999) have exploded the old notion that the level of Euro-Indian trade was *de minimis*, which rested on erroneous views of American Indians as culturally uninterested in market exchange, devoid of monetary and credit instruments, and possessing indifferent farming skills at best (Swanson 2018, 10–32). In fact, many Indians were entrepreneurial (Miller 2012), ranched salmon (Johnsen 2009), farmed extensively, maintained ceramic, food, and lithic tool manufacturing sites (Stothers and Abel 1993), extended credit (Johnsen 2016), and used their own forms of money (well beyond wampum) as well as European monies (Wright 2019, 148–66), from coins to paper money (Millan 1744, 11). Many held very strong views on the sanctity of private property and only began to steal when oppressed by Euroamerican laws and troops (Cooper 1929, 286). As one Hudson Bay adventurer put it, Indians "are bold and enterprising even to enthusiasm, whilst there is a probability of success crowning their endeavours; but wise enough to desist, when inevitable destruction stares them in the face" (Wales 1770, 110).

By the late seventeenth century, many Indian women were each brain tanning eight to ten deer hides a day for domestic consumption and trade with Virginians and Carolinians (Southwell 1686–1692, 532–33). By that time, French Louisiana also traded heavily for deerskins (Usner 1985). Later, Georgia became a major deerskin trade center as well (Pressly 2013, 198–200). Overall, trade in deerskins in North America amounted to hundreds of thousands per year in the mid-eighteenth century (Usner 1992, 246; Dunaway 1994, 225).

This prodigious harvest Indians managed, without the aid of horses, through the use of live decoys, their own bodies adorned with fake antlers while pretend foraging (Clayton 1694, 121–22), as well as drives using fire, noise, and natural barriers and corrals (Waselkov 1978, 18). Fire drives continued in some parts of the country through the end of the nineteenth century (Anon. 1883c; Disagreeable Experience 1892) and were so common earlier that there is evidence that they changed forest composition, at least in the southeast (Foster and Cohen 2007). Indian hunters also ran deer for hours, on foot, until their quarry was too exhausted to flee further (New York Times 1885).

Nevertheless, deer remained so abundant throughout the seventeenth century that "a good Woodman [a colonist living away from settled areas] ... will keep a House with Venison [meet his family's protein needs with deer meat]" (Clayton 1694, 122). Most colonists, though, subsisted mainly on hogs, chickens that did not fall prey to "Rackoone," and small wild game like rabbits and squirrels (Clayton 1694, 122–23).

It is likely that the white-tailed deer-carrying capacity of the land increased over the colonial period due to increased agriculture. Decreased natural predation may have played a role as well, as wolves were said to prefer sheep to the point "that a piece of Mutton is a finer Treat, than either Venison, Wild Goose, Duck, Widgeon, or Teal" (Clayton 1694, 122, 125; Savage 1825, 34, 59–60, 115). A literal "lone wolf" in the desert southwest in the early twentieth century was said to have killed $25,000 worth of livestock, mostly sheep and cattle calves, before government hunters gunned him down (Anon. 1925b). Bears, by contrast, seemed to prefer pork (Wright 1868a, 121). To some extent, however, open-range pigs and cattle competed with deer for hard mast (tree nuts) and browse near settlements (Reitz and Waselkov 2015, 24; Swanson 2018, 29).

In Maryland, clearing the forest, which observers said covered everywhere, was sped by the desire to plant tobacco and other crops, and the need for barrel staves. Despite rising human populations, "Deer, Fowle, both Water and Land, [remained] in abundance" throughout the seventeenth century (Jones 1699, 436–37). The few Indians remaining in eastern Maryland added to the harvest and "took delight in nothing else" (Jones 1699, 441).

Yet the deer population did not experience precipitous decline over the eighteenth century. Bartram in the mid-1760s speaks of his small party harvesting deer, turkeys, and other game animals without difficulty during his tour through Florida, Georgia, and the Carolinas (Bartram and Harper 1942). When John Davis, a recent émigré from England, lost his job teaching at a college in South Carolina in the 1810s, a planter named

Brisbane "invited me to his plantation, to partake with him and his neighbors, the diversion of hunting during the winter" (1817, 16). Davis found that "the woods abound with deer, the hunting of which forms the chief diversion of the planters," who used dogs and slaves to drive deer toward posted shooters. Wolves and cougars still inhabited the area but, as centuries earlier, often found livestock easier pickings than deer, which Davis reported showed little fear of humans, suggesting that they were not often pressured (1817, 17) because "the deer is by nature a timid animal, and persecution makes it more so" (Wright 1868b, 466). In 1838, Audubon spoke of the "incredible abundance" of America's white-tailed deer but warned that under current hunting techniques, they could be as rare as cervid species in Britain "in a few centuries" at their then rate of decline.

It appears that as the human population increased over the nineteenth century, natural predation decreased and carrying capacity increased enough to offset increased human predation. Even in relatively densely populated Long Island, the deer population grew right alongside the human one (New York Commercial 1841). Vehicle deaths were minimal because the main threat, trains, ran relatively infrequently, especially at times when deer are most active. In fact, deer inadvertently run over by trains were considered newsworthy in antebellum America (Anon. 1844). Restrictions placed on hunting in the colonial and early national period, which ranged from closed seasons to land ownership requirements to bans on driving deer with fire, probably did nothing more than reduce incentives to engage in commercial hunting, but that appears to have been sufficient to maintain population stability (Smalley 2016, 314–16).

The early postbellum period was generally a good one for game in the South as abandoned farms and orchards provided cover and food for a variety of species large and small (Ferril 1893). By the Gilded Age, however, deer were "fast disappearing" from "the States east of the Mississippi," where they were hunted for their meat and skins as well as sport (Wright 1868b, 466, 474). Every deer would be killed, contemporaries believed, if they were as easy to hunt as bison (O. 1889). Some claim that by the 1880s, white-tailed deer had been extirpated from all of New York State outside of the thinly populated and difficult to traverse Adirondack Mountains (Braverman 2015, 158 n.99). Most of upstate New York's thick woodlands had been clear cut, which caused high winter mortality due to a dearth of cover and browse when the snows ran deep, as they often did (Dickinson 1993, 75–76). In the 1890s, however, deer could still be found in "thick woodlands" remaining on more temperate Long Island even though they fetched $12 apiece at market (Anon. 1895c).

Deer also remained numerous in backwoods Pennsylvania, even though their meat sold for 8 to 13 cents per pound (Erie Dispatch 1884; Clemens 1886). And in Michigan and Wisconsin, deer were said to be nearly as numerous as ever even while "many men make a living" killing and selling them (New York Mail and Express 1886).

White-tailed deer populations also stayed strong west of the Mississippi, especially Texas, where by the 1870s more livestock grazing and fewer fires helped to spread whitetail-friendly juniper (Bryant 1991, 60–61). Whitetails were also considered abundant in southwest Missouri and Arkansas in the 1870s (Anon. 1876b), though subject to the "black tongue" (Anon. 1872) disease (likely bluetongue, an orbivirus that affects deer populations [Kluender et al. 1992]). In the 1880s, white-tailed deer were still thought "in fair abundance" along the Cimarron River in Oklahoma (Hough 1889, 428). The mountains of Tennessee were then said to be "alive with deer, and there is plenty of venison" at market (Anon. 1881d). Following the Panic of 1893, Colorado miners said they could live on a "straight meat diet" composed of "deer and elk meat" until better prices for ores could be had (Anon. 1893c). In the late 1890s, 100,000 white-tailed were said to inhabit the woods of Maine, which was allegedly "more than in all the other New England and middle States combined" (Anon. 1899b). With about 25,000 harvested per year, "no extinction of deer is apt to occur" if game laws were enforced according to famous woodsman Jack Darling (Anon. 1898c). In that spirit, California placed various moratoria on the shooting of blacktail (*Odocoileus hemionus columbianus*) in the 1870s (Anon. 1870c, 1878).

In Michigan, white-tailed deer numbers appear to have kept up because many areas were inaccessible to hunters due to a lack of roads or legal permission to hunt. Deer remained "very plentiful in some localities" in Michigan well into the 1890s (Chicago Record 1896). Tellingly, it was called a "slaughter-yard" when 50 or 60 deer a day were harvested in southeastern Michigan in 1854 (Toledo Rep.) and it was considered a mass slaughter when hunters killed 70,000 deer a year in Michigan in the 1880s (W. S. 1924, 365). Recently, though, Michigan hunters have harvested 300,000 plus deer a year without damaging a population estimated at 1.75 million (Mack 2017).

Similarly, people in Wisconsin feared that deer would be exterminated when only 6000 were killed for market (Anon. 1866e, 1898d) in a state where the annual harvest is now 220,000 out of a herd of over 1.5 million (Smith 2019). In Maine in the 1870s, a harvest of 42 deer was offered as evidence that "a large number" of deer were "falling victim" to hunters (Bangor Whig 1873). Today, about 300,000 deer inhabit Maine, where the annual harvest is about 30,000 (Holyoke 2019). The harvest of 2000 deer by "Sioux Indians" in Minnesota in 1860 was enough to upset the "poor whites ... who are fond of venison" (Anon.). In 2019, Minnesota hunters harvested almost 182,000 deer out of an estimated population of 1 million (https://www.dnr.state.mn.us/mammals/deer/management/statistics.html). Carrying capacity may have increased in those places in the interim but not by an order of magnitude, suggesting that hunter harvest in the nineteenth century was underestimated, or, if the harvest estimates mentioned above were accurate, well shy of exceeding sustainable levels.

By 1900, people noticed white-tailed deer were beginning to stay around settled areas, even within city limits (Anon. 1899b). An estimated 300,000 white-tailed deer then remained nationwide, many in urban and other enclaves (Zink 2014, 6), but the actual number will probably never be known.

Small Game

Despite urban markets for wild ducks, rabbits, and squirrels throughout the antebellum period in places like Chillicothe, Ohio (Anon. 1849), and elsewhere (see Fig. 3.1), small game seems to have remained abundant, if only in areas where people were too abundant to hunt or in largely inaccessible backwoods and swamps (Anon. 1883a; Vicksburg Commercial Herald 1892). Upland game birds remained plentiful in Missouri in 1881 (Anon. 1881e) and in Oregon in 1894, despite being sold to markets in strings of 50 (Anon. 1894c).

When and where the road system was less developed, like the antebellum South, small game numbers seemed to have kept up especially well.

GAME – VENISON 9 @ 12¼ ₱ LB; HARE, $4 @ $4; RABBITS, $4.50; SNIPE, $3 FOR ENGLISH AND $1 @ 1.25 FOR COMMON; PLOVER, CURLEW, DOVES AND LARKS, $1 @ 1.50 ₱ DOZEN

Fig. 3.1 Market prices for wild game meat, 1878. (Redrawn from original source: Anon. 1878)

Davis (1817) reported downing doves (*Zenaida macroura*), geese (*Branta canadensis*), turkeys, and squirrels in prodigious numbers. In New Hampshire in 1819, ten men claimed to have harvested *in a single day* 1665 "grey and red squirrels, heath-cocks, hawks, owls, ducks, partridges, crows, rabbits, muskrats, minks, hedge-hogs, foxes, and otters" (Anon. 1819).

Even in relatively densely populated Connecticut, squirrel hunters had no complaints about a dearth of targets in the 1880s (New York Mail and Express 1886). Loss of habitat led to a decrease in the number of gray squirrels (*Sciurus carolinensis*) in southwestern Missouri in the late nineteenth century. Fox squirrel (*Sciurus niger*) numbers in the same region stayed up, though, because they were more tolerant of humans and smaller trees (Basket 1893).

The spread of roads, combined with the American shoot-on-sight attitude, eventually led to the diminution of various game birds and mammals in Minnesota. Only well-heeled hunters, some from the East and England and others from Midwestern cities like Chicago and Milwaukee (Anon. 1895b), could afford to hire special train cars outfitted for hunting expeditions that lasted a month or more. Many locals, however, outfitted hunting wagons to support short trips into the most productive hunting regions (Berthel 1935, 260, 269–70).

Some species subject to commercialization seem to have suffered from reduced ranges but not necessarily because markets for their meat existed. Edwards (1753–1754, 500) reported, for example, that "Pennsylvania pheasants" (undoubtedly the ruffed grouse, now the state bird, *Bonasa umbellus*), "have been common (in Pensylvania [*sic*]), but now most of them are destroyed in the lower settlements, tho' the back Indian inhabitants bring them to market." Undoubtedly, changing habitat played a key part of the declining population in the urbanized and farmed parts of the province because grouse were absent only from the treeless parts of Maryland (501).

In Maine, by the end of the nineteenth century, snipe and woodcock were to be found only "where hunters seldom go" (Anon. 1899b). Unsurprisingly, other game animals, including "partridges," were also most abundant where the fewest hunters tread (Anon. 1891c). Small game also remained plentiful in remote areas of the Ozarks (Anon. 1876b).

Waterfowl

The number of Canada geese (Jonker et al. 2006, 1009) and other migratory waterfowl game species fell dramatically in the late nineteenth and early twentieth centuries, due more to the use of nets and traps (Ferril 1893; Boston Journal 1886), four bore shotguns capable of dropping 30–60 geese per shot (Anon. 1879), and lead pellets (Detroit Free Press 1892; Anon. 1932) than sport hunting pressure. Fowling nets were hundreds of yards long and could capture 800 ducks twice daily. When employed by competing groups of commercial hunters, they decimated entire flocks in a few days (Baltimore American 1887).

In 1870, one commentator argued that the "enormous prices" offered for canvasback ducks would be the "cause of their rapid extermination," though he admitted that "a decrease in their numbers is now imperceptible" (Harper's Magazine). When hunters near Milwaukee bagged only 1–5 ducks a day, instead of the 30 plus they were accustomed to, they were "discouraged" (Anon. 1895b). The death of 100 ducks, apparently habituated to humans, at the hands of a few hunters in a few hours made the newspapers in 1883 (Baltimore Sun). Banning trapping and moving to non-toxic shot, rather than the ending of wild waterfowl markets per se, sparked the rebound of ducks and geese. It also helped that farmers met market demand with a more uniform product not laden with lead.

Meleagris Gallopavo

Many species of wild game were saved, locally if not globally, by making them private property on Indian Reservations (Reagan 1919–1921, 444) and game reserves, which in America date to the late nineteenth century (Crossways 1896). The Morris game reserve in Hammond, Louisiana, for example, became a major refuge for turkeys around the turn of the twentieth century under the private management of Charles L. Jordan (W. S. 1915, 115–16). Once common on the bayou, turkeys were extirpated in many swampy areas of the state in the late nineteenth century but remained numerous on reserves (Brasseaux et al. 2004, 81).

While in the late nineteenth century there was some sense that turkeys were not as plentiful as previously (Taylor 1888), in the early 1880s turkeys were still considered in "great abundance" in parts of Pennsylvania, though some thought them destined for extinction due to commercial hunting to feed the markets in Philadelphia, New York (Anon. 1882), and hundreds of lesser towns like Terre Haute, Indiana (Taylor 1888). In the late 1880s, though, turkeys were still to be found in good numbers in the rural districts of the Midwest and said to be "beautifully plenty within 15 miles of Harrisburg" (Williams 1889). A decade later it was said that the forests surrounding the James River plantation of H. B. Bilt in Virginia abounded in "deer, wild turkeys, and smaller game" (Malley 1895). Turkeys, however, were apparently exterminated in Michigan by the 1920s (W. S. 1924, 365). They were baited and trapped or hunted with horses and dogs, or over bait, or with gobbler and hen calls (Anon. 1882; Taylor 1888; Cor. New York Herald 1889; New York Sun 1889). Hunters even shot them off their roosts at night (Taylor 1888).

Again, though, farmers eventually met demand with a more uniform product not laden with lead. Shrinking supply and higher prices produced incentives to find inexpensive ways to farm them.

Ursus *Sp.*

American Indians and Euroamericans long sought black bears for their hides, fat, and meat, which they described as being "betwixt Beef and Pork." Cubs were especially esteemed for their taste (Hairr 2011, 320–21). Many bear "hunters" were actually trappers who concocted all sorts of bruin traps out of steel and springs and wood and holes (Weyawega Times 1872; Lewiston Journal 1885; Mighels 1897; King 1899). When hunted, it was often with dogs and/or at night (Anon. 1889). Some bear hunters bragged of using exploding bullets (Exchange 1888).

In the 1850s, black bears, despite being harvested for market, were abundant enough in the Appalachians that there were men who still devoted "their whole existence to bear hunting." Although bears were not as plentiful as "in the days of Boone, ... the hunt is tolerably productive to the persevering hunter" (Anon. 1854a, 99). Bears in the mountainous parts of California remained "thick as ever" through the 1880s (San Francisco Examiner 1887), when bears in Pennsylvania's mountains also remained plentiful, even though they killed livestock and their dressed carcasses brought from 5 to 8 cents a pound at market (New York Sun 1884; Altoona Tribune 1886; Clemens 1886). Harvesting large bears was so lucrative that hunters were known to crawl into their dens to kill them (Anon. 1876a; King 1899). Yet despite their relatively slow reproductive cycle, bears remained "too thick for good neighborhood" throughout most of Appalachia (Williams 1889). Bears remained abundant in Idaho in the 1880s as well (San Francisco Examiner 1888).

While black bear range has shrunk by approximately 40 percent from its historical apex (Laliberte and Ripple 2004, 126), their numbers are generally stable or slowly growing where they are still to be found, though population estimates based on bear nuisance activity, vehicle collisions, sign surveys, and hunter questionnaires, while good enough for management purposes, are not "precise or rigorous enough to provide useful information on population trend" (Garshells and Hristienko 2006).

The range of *Ursus arctos* (brown bears or "grizzlies") also shrank though they were killed mostly for sport, their hides, or to safeguard livestock. They, too, were hunted with dogs, at night, and such (Portland Oregonian 1883) and also trapped with giant steel traps anchored with logging chains and baited with live mice (New York Times 1887).

Large Ungulates

Bison, wapiti, and moose range retreated from the eastern states to the West and North by the 1840s, a few doomed strays excepted (Anon. 1842, 352; London Saturday Review 1887; Branch 1929). Areas at the edges of their ranges near urban areas or transportation routes suffered heavy hunting pressure exacerbated by markets. Wapiti, for example, were prized for their hides and their fat because of "its superior whiteness, hardness and delicate taste" (Anon. 1893a).

Moose hides were long known to produce machinery belts every bit as good as bison hides but moose were prohibitively expensive to hunt due to their swampy, wooded, isolated habitat and solitary social habits (Dudley and Chamberlayne 1720–1721). Nevertheless, by 1879 there were in some places "more callers than moose" as one Indian guide put it (Dunraven 1879, 343, 352). One observer in Maine claimed in 1884 that moose were "now nearly extinct" in the state (New York Times 1884). In 1891 it was claimed there was not a single moose within 200 miles of Ottawa and Ontario banned moose hunting entirely for years on end, though antler poaching remained common (Anon. 1891b). By 1899, one Maine observer was confident that with 15,000 hunters taking only 2000 moose, mostly young bulls, "the moose is following the buffalo [sic] on the road to utter extermination" (Anon. 1899b). Others also believed the moose was doomed to extinction (London Saturday Review 1887) because they conflated extinction with range shrinkage.

North and West maintained wild spaces, especially at higher elevations, large and remote enough to harbor sizable numbers of both moose and wapiti (Dunraven 1879, 339; Laliberte and Ripple 2004, 132) until implementation of the NAWCM. Parts of the great north remained so remote, after all, that reports of living mammoth were made (Forest and Stream 1896). The veracity of such reports seems unlikely but clearly vast uninhabited stretches ensured reservoirs from which depleted species could rebound. Dall sheep numbers were way down in Alaska in the early twentieth century, for example, because miners had killed at least 10,000 of the delicious creatures (by all accounts the best tasting meat in North America, with the possible exception of mountain lion) during the various gold and other metal rushes (Rockwell 1923). Yet they survived the carnage. Miners also killed enough moose to supposedly threaten local populations, yet a hunter was able to take two mature bulls, a young bull, and a cow in about an hour (Rockwell 1924).

By 1900, moose, like white-tailed deer, learned to live around settled areas, even within city limits (Anon. 1899b). One hypothesis is that the NAWCM prevented people from shooting game animals off their back porches whenever they were spotted (New York Mail and Express 1886; Johnston et al. 1896), allowing habituation to occur, especially in urban "no hunting" enclaves. (Raccoons, foxes, and other species are also becoming commensal, like mice and rats. The benefits of living in close proximity to humans outweigh the costs when human predation decreases [Budiansky 1992, 45–52]).

In the West, moose range may have spread south after the Indian population shrank but before pressure from Euroamerican hunters increased (Kay 1997). They eventually moved back into Yellowstone's Northern Range, for example, albeit at low density (Despain et al. 1986, 52–53). Caribou, once found in northern Minnesota and Maine, shrank away to the north (Anon. 1891c, 1898c, 1899b; Reagan 1919–1921, 444; Berthel 1935, 261). In the late 1880s wapiti were still found "in abundance" in Idaho, which boasted of having "more deer" (mostly mule deer) than any other place "on all the earth" (San Francisco Examiner 1888).

Today, the range of the wapiti has been reduced by approximately three-quarters, while moose range has shrunk only about ten percent. Sophisticated statistical models based on scientific surveys and radio collaring help wildlife managers in Alaska to predict the population trends of moose, which remain important sources of protein for locals (DeLong and Taras 2009).

Caribou range decreased about a quarter from its historical height (Laliberte and Ripple 2004, 126). Climate change, rather than human predation, appears to be the most important driver of late. Global climate change will increase the need for high-quality scientific wildlife management in the future (Biber and Eagle 2015, 787–88, 800).

Bison Bison

The range of bison is only about one percent (Laliberte and Ripple 2004, 126) of its impressive apex, which extended from Alaska to the southern Louisiana bayou (Brasseaux et al. 2004, 81). While herding and flocking can serve to protect individual members of a species, huge assemblages of animals in open country subject them to devastating market forces. Both bison and passenger pigeons were especially vulnerable to human predation because they accumulated in large groups that kept the cost of hunting them relatively low even when their overall numbers were declining. Most other animals become more difficult/costly to locate as their numbers shrink, leading to a boom-bust population cycle, as with rabbits and grouse, rather than a population crash (Speck and Eiseley 1942, 220). Markets drove bison to the brink of extinction but also ultimately saved by them (Isenberg 2000).

Bisons were under stress even before Euroamericans arrived on the Great Plains as Indians using rifles and horses tapped the common pool to supply markets for bison hides (Baden et al. 1981; Isenberg 2000). Demand for bison hides, in the summer for industrial belting and in the winter for "robes," created incentives to harvest them (New York Sun 1878). By 1860, their range was already shrinking to the west but contemporaries believed that supplies at market decreased only because Indians, then their most avid hunters, were going extinct, "not from any decrease in the buffalo [sic]" (New York Commercial Advertiser 1860). Most of their meat fed scavengers and was sometimes used to poison them (Miles 1895, 492). Pioneers years later collected and sold the bleached bones (Anon. 1884b), which were shipped east to be processed into fertilizer (Branch 1929).

But the mere existence of demand was insufficient to cause their near extinction. It was the relative ease of killing them, the indiscriminate nature of the slaughter, and habitat loss that almost led to their extermination (Anon. 1873a; Dunraven 1879, 360–61). Railroads hurt them coming and going, first by feeding the construction workers that built the lines and then by decreasing the cost of transporting their hides to eastern factories, where they were cut into belts and tethered to steam engines that drove machinery (San Francisco Examiner 1888; Miles 1895, 490; Braverman 2015, 145–46).

"Never before in all history," one newspaper noted, "were so many large wild animals of one species slain in so short a space of time," about a dozen years (Anon. 1886d). Government aided and abetted the destruction, hoping it would end the Indian military threat and open the Great Plains to settlement and cattle ranching (Isenberg 2000).

Markets for bison body parts, however, ultimately saved the species by giving people an incentive to preserve them on private ranches once their numbers on the common open prairie reached a low ebb (Western Hunter 1889). Although naturally migratory, they survived anywhere sufficient fodder and water could be found, including New Hampshire (Anon. 1895c). Unfortunately, many were cross-bred with cattle. Today, many privately owned herds, some genetically pure, are thriving and publicly owned herds in South Dakota, Wyoming, and Montana support limited hunting (Isenberg 2000).

Ectopistes Migratorius

According to Williams (1889), "nature has so intrenched herself in rock and mountain fastnesses that 500 years of hunting will not exterminate game there." That may be, but bison could not live in such crevices and neither could passenger pigeons, which succumbed to markets, habitat loss, bad timing, a plethora of close substitutes, and the difficulty of domesticating them. Passenger pigeons once numbered in the billions, ironically perhaps due to an unintentional but nevertheless manmade increase in mast, their primary food source (Neumann 1985). In the 1850s, they were so numerous that Indiana farmers considered them a nuisance akin to locusts (Anon. 1854b). In the early 1880s, their largest breeding roosts were 20 miles long and 15 miles across (Anon. 1881c).

Whatever the cause of the passenger pigeon's efflorescence, humans killed them in large numbers with incredible ease. Every part of their tiny bodies, including their feces, which was thought to have medicinal properties, found ready markets both before and after the arrival of Columbus in the New World (Wales 1770, 117). They also contained a highly prized oil (Neumann 1985, 399, 404). Habitat loss and improved netting (a party of five could net 400 dozen pigeons a week) and shooting technologies (three score could be killed with a single shotgun shell) increased the pressure (Anon. 1869). Small groups of hunters were known to harvest thousands in just a few days (Memphis Avalanche 1883) and ship them east by the barrel (Winona Republican 1873). Trappers took them in their breeding grounds, keeping some alive to supply the sport shooting trade in the East, where the fast, agile little birds were esteemed much more difficult to down than domesticated rock pigeons (Anon. 1881c).

Despite their market value, passenger pigeons, unlike bison, mink, and many other animals, were not privatized because they did not breed well in captivity, perhaps due to underfeeding or inappropriate diet (Neumann 1985, 397). Little incentive to discover techniques for domesticating them existed because close substitutes had long since been domesticated (Millan 1744, 34, 39). In addition to being harvested commercially and subject to pest eradication hunts (Neumann 1985, 397), passenger pigeons were shot for sport along with other types of *Columbidae* (pigeons and doves) that did not flock in such huge numbers. In the early twentieth century, hunters harvested remnant groups, mistaking them for mourning doves, but the fate of this gregarious bird had been determined in the 1880s and 1890s, when hunters disrupted their last few remaining breeding grounds. NAWCM rules came too little, too late for these little fast flyers, which would have benefitted from bag limits and harvest restrictions, especially given that breeding habitat loss, their great flocks, and modern communication technology made it too easy to locate and disturb their increasingly limited breeding range (Clemens 1886).

Conuropsis Carolinensis

Markets also played a role in the 1910 extinction of the Carolina parakeet, which was hunted mostly for its feathers for the millinery (fancy hat) trade and sometimes as an agricultural pest (Hairr 2011, 314). Deforestation may have stressed the species further but the cause of its rapid decline early in the twentieth century remains a mystery. A pathogen may have provided the final shock. With little warning of its demise, no attempt to domesticate it appears to have been made.

To Hunt or to Ranch, That Is the Question

Privatization/domestication efforts rely on both economic and biological conditions. Colonists introduced managed honeybees (*Apis mellifera*) into parts of New England, for example, but subsequently allowed them to re-wild when they discovered an easy method for "hunting" them (discovering their hives) using bait stations and simple triangulation (Dudley and Chamberlayne 1720–1721, 148–50). Loss of that technique for reasons unknown, plus the rise of markets for honey and then pollination services, however, tipped the scales back toward beekeeping, though dabblers continued to "bee hunt" for personal consumption each fall (Anon. 1881b; Baraboo 1899; Pellet 1938). Markets for pollination services provided beekeepers with strong incentives to mitigate the effects of bee diseases, including Colony Collapse Disorder, and they quickly did so (Rucker et al. 2019).

Similarly, many North Americans found it cheaper to turn their hogs and even cattle loose to forage for themselves rather than to actively ranch them (Reitz and Waselkov 2015, 36; Swanson 2018, 29). When needed, they were trapped or hunted, though not without some cost and danger because they quickly reverted to their wild forms and temperaments (S.H. 1833; Knapp 1935, 84). Many frontier farmers did not raise any livestock because hunting yielded as much meat with much less toil or trouble (Smalley 2016, 324).

Ruffed grouse are infamously difficult to hunt after they learn to fear humans and many people find their flesh delicious. So a strong incentive to domesticate them existed but early attempts to farm them failed because "as soon as hatched, they run into the woods" (Edwards 1753–1754, 500). Similar problems were encountered with turkeys (Anon. 1842, 358), but better caging and other technologies and techniques (e.g., to keep males from destroying eggs [Sporting Journal 1885]) eventually allowed for their farming, and even pheasants (*Phasianus colchicus*) and other upland game birds (McCarty 1934). Farmed game birds can be sold for human consumption or used to stock hunting grounds, though not without controversy.

References

A Hunter. 1881. A Hunting Party. *Arkansas Gazette* (31 Dec.)
Altoona Tribune. 1886. A Hunter Rescued by His Dog. *St. Louis Globe-Democrat* (21 Dec.), 8.

Anderson, Terry L. 1998. Viewing Wildlife Through Coase-Colored Glasses. In *Who Owns the Environment?* ed. Peter J. Hill and Roger E. Meiners, 259–282. New York: Rowman & Littlefield.

Anderson, Terry L., and Gary D. Libecap. 2014. *Environmental Markets: A Property Rights Approach.* New York: Cambridge University Press.

Anon. 1819. Great Hunting. *National Intelligencer* (25 Nov.).

———. 1842. Walton and Cotton's Complete Angler.... *North American Review* 55 (117): 343–372.

———. 1844. Railroad Venison. *National Intelligencer* (27 Nov.).

———. 1849. City Brickbats and Pebbles, Picked Up in the Streets. *Daily Scioto Gazette* (22 Dec.).

———. 1854a. Bear Hunting. *The Illustrated Magazine of Art* 4 (20): 99–100.

———. 1854b. Wild Pigeons. *Daily South Carolinian* (13 Oct.).

———. 1866a. Thirty Years of Army Life ... by R. B. Marcy. *North American Review* 103 (213): 80–82.

———. 1866b. The Marten Trappers. *Frank Leslie's Illustrated Newspaper* (13 Oct.), 60.

———. 1866c. Trapping Martens in Canada. *Frank Leslie's Illustrated Newspaper* (8 Dec.), 188.

———. 1866d. Trapping the Beaver. *Daily Evening Bulletin* (29 May).

———. 1866e. We Have Heard an Estimate. *Milwaukee Daily Sentinel* (18 Jan.).

———. 1867. A Very Homely But Readable Book. *Frank Leslie's Illustrated Newspaper* (14 Dec.), 194.

———. 1869. Pigeon Trapping. *Wisconsin State Register* (24 April).

———. 1870a. Wild Life in the West. *Milwaukee Daily Sentinel* (10 Feb.).

———. 1870b. Indian Trappers. *Daily Evening Bulletin* (21 Feb.).

———. 1870c. Venison in Market. *Daily Evening Bulletin* (2 July).

———. 1872. The St. Paul Press Is Informed. *Milwaukee Daily Journal* (13 Nov.).

———. 1873a. Buffalo Hunting. *Daily Graphic* (21 July), 134.

———. 1873b. The Fur Trade of Lewiston. *Lowell Daily Citizen* (23 Dec.).

———. 1876a. A Bear Hunting Exploit. *St. Louis Globe-Democrat* (13 Aug.), 12.

———. 1876b. Hunting and Fishing. *St. Louis Globe-Democrat* (209 Oct.), 6.

———. 1876c. There Appears to Be Quite a Rivalry. *Arizona Miner* (29 Dec.).

———. 1877. A New Hunting Ground. *Scientific American* 36 (4): 49.

———. 1878. Contraband Venison. *Daily Evening Bulletin* (3 Aug.).

———. 1879. Hunting Wild Geese with Oxen. *Daily Inter Ocean* (8 Dec.), 6.

———. 1881a. Hunting Alligators in Florida. *Scientific American* 45 (18): 297.

———. 1881b. Hunting Wild Honey. *Frank Leslie's Illustrated Newspaper* (1 Oct.), 71.

———. 1881c. Wild Pigeons for the Sportsmen's Tournament. *Frank Leslie's Illustrated Newspaper* (2 July), 299.

———. 1881d. The Mountains of Tennessee. *News and Observer* (1 Jan.).

———. 1881e. Preparing to Sacrifice Game. St. Louis Globe-Democrat (14 Oct.), 7.

———. 1882. Hunting Wild Turkeys. *Daily Inter Ocean* (9 Feb.), 5.

———. 1883a. Hunting Stories. *St. Louis Globe-Democrat* (15 Dec.), 16.

———. 1883b. The Bellefontaine Trappers. *St. Louis Globe-Democrat* (17 Feb.), 9.

———. 1883c. At Red Bluff the Sky Is Clouded. *Los Angeles Times* (6 Sept.).

———. 1884a. Trapping in Wisconsin. *St. Louis Globe-Democrat* (17 Oct.), 4.

———. 1884b. Report Says. *Galveston Daily News* (23 Nov.), 3.

———. 1885a. Fish Trapping. *Idaho Avalanche* (4 Apr.).

———. 1885b. A Muskrat's Perils. *Wisconsin State Register* (26 Dec.).

———. 1886a. Deer Hunting with Steam. *Scientific American* 55 (23): 354.

———. 1886b. Two Hunting Stories. *Detroit Free Press* (15 May), 8.

———. 1886c. Going Fox-Hunting. *Daily Inter Ocean* (11 Nov.), 3.

———. 1886d. His Hunting Trips. *Milwaukee Daily Sentinel* (2 Apr.), 4.

———. 1886e. D.P. Graves, the Well-Known Hunter. *Milwaukee Daily Sentinel* (29 Nov.), 8.

———. 1888. A Border Trapper's Life. *Rocky Mountain News* (11 Nov.), 23.

———. 1889. Hunting Bear in Louisiana. *Daily Inter Ocean* (26 May), 12.

———. 1890. Some Tall Hunting Stories. *Milwaukee Journal* (25 Oct.), 6.

———. 1891a. Squirrel Hunting. *Atchison Daily Champion* (16 Sept.), 2.

———. 1891b. Hunting the Wary Moose. *Daily Inter Ocean* (29 Nov.), 29.

———. 1891c. Hunting Adventures. *Bangor Daily Whig and Courier* (28 Oct.).

———. 1893a. A Mighty Hunter from the South. *Daily Picayune* (9 Sept.).

———. 1893b. Trapping Beaver. *Boston Daily Advertiser* (23 Mar.), 8.

———. 1893c. Will Live on Venison. *Rocky Mountain News* (9 Sept.), 3.

———. 1894a. The Mightiest Hunter. *Atchison Globe* (27 Apr.), 3.

———. 1894b. Hunting Horses. *Vermont Watchman* (9 May), 2.

———. 1894c. Markets Full of Game. *Portland Oregonian* (31 Oct.), 5.

———. 1895a. Hunting Egrets in Mexico. *Daily Inter Ocean* (2 Dec.), 3.

———. 1895b. Hunters and Hunting. *Milwaukee Daily Sentinel* (3 Nov.), 11.

———. 1895c. The Hunting Paradises. *Daily Picayune* (29 Sept.), 21.

———. 1895d. Private Secretary Thurber's Zeal. *Daily Inter Ocean* (17 June), 6.

———. 1895e. Moose Hunting. *Bangor Daily Whig and Courier* (3 Dec.).

———. 1895f. Is a Brave Hunter. *Daily Inter Ocean* (28 Nov.), 7.

———. 1895g. Antelope Trapping in the West. *Daily Picayune* (2 Nov.), 12.

———. 1895h. Muskrats Are Sly Animals. *Emporia Daily Gazette* (1 March).

———. 1896a. Duck Hunting at Home. *Milwaukee Journal* (27. Nov.).

———. 1896b. A Chicago Hunter. *Galveston Daily News* (4 Dec.), 8.

———. 1896c. Trapping in Maine. *Daily Inter Ocean* (17 Apr.), 10.

———. 1896d. Raccoons in Louisiana. *Portland Oregonian* (19 Jan.), 15.

———. 1897a. Flamingo Hunting. *New Orleans Daily Picayune* (16 Aug.), 7.

———. 1897b. In Conversation. *Bangor Daily Whig and Courier* (16 Oct.)

———. 1898a. Danger in New Hunting Rifles. *Scientific American* 79 (6): 85.

———. 1898b. Hunting and Fishing. *Milwaukee Journal* (4 Nov.), 4.

———. 1898c. Hunting in Maine. *Atchison Globe* (5 Oct. 1898), 3.

———. 1898d. Sold Confiscated Venison. *Milwaukee Daily Sentinel* (27 Nov.), 12.

———. 1899a. Coon Hunting. *Bangor Whig and Courier* (13 Oct.), 3.

———. 1899b. The Hunting Season. *Bangor Daily Whig and Courier* (20 Oct.), 8.

———. 1899c. A Woman Bear Hunter. *Boston Daily Advertiser* (13 June), 8.

———. 1901. Fall Hunting in the Maine Woods. *Journal of Education* 54 (15): 246.

———. 1902. The Hunting Season Now at Its Height—Deer and Moose Very Plentiful This Year in Maine. *Journal of Education* 56 (17): 291.

———. 1913. Fox-Hunting in America. *Lotus Magazine* 5 (1): 51–67.

———. 1918. Camouflage in Seal Hunting. *Scientific American* 118 (17): 383.

———. 1923. My Life with the Eskimo.... *Geographical Teacher* 12 (3): 218–219.

———. 1925a. Handbook of Alaska.... *Washington Historical Quarterly* 16 (4): 306–308.

———. 1925b. Notorious Wolf Killed by Government Hunters. *Science News-Letter* 6 (203): 6.

———. 1932. Lead Shot Kills Ducks Even Though Hunters Fire and Miss. *Science News-Letter* 21 (566): 107.

———. 1899. Good Muskrat Trapping. *Atchison Globe* (10 Nov.), 3.

Archibald, Malcolm. 2013. *The Dundee Whaling Fleet: Ships, Masters and Men.* Edinburgh: Edinburgh University Press.

Audobon's Ornithological Biography. 1838. Deer Hunting. *National Intelligencer* (10 Sept.).

Baden, John A., Richard Stroup, and Walter Thurman. 1981. Myths, Admonitions, and Reality: The American Indian as a Resource Manager. *Economic Inquiry* 19 (1): 132–143.

Baltimore American. 1887. Wild Duck Trapping. *St. Louis Globe-Democrat* (5 Nov.), 15.

Baltimore Sun. 1892. A Cruel Method of Hunting. *Bismarck Tribune* (24 Jan.), 3.

Bangor Whig. 1873. Large Numbers of Deer. *Lowell Daily Citizen* (31 Jan.).

Baraboo, Wisconsin. 1899. Hunting the Wild Bee. *Milwaukee Daily Sentinel* (15 Jan.), 11.

Barboza, Perry S., and Daniel Tihanyi. 2018. State Wildlife Policy in a National Environment. In *North American Wildlife Policy and Law*, ed. Bruce D. Leopold, Winifred B. Kessler, and James L. Cummins. Boone and Crockett Club: Missoula.

Barker, John. 1896. Frozen to Death While Out Hunting. *New Orleans Daily Picayune* (30 Nov.).

Barnes, Jonathan I., James Macgregor, and L. Chris Weaver. 2002. Economic Efficiency and Incentives for Change with Namibia's Community Wildlife Use Initiatives. *World Development* 30 (4): 667–681.

Bartram, John, and Francis Harper. 1942. Diary of a Journey Through the Carolinas, Georgia, and Florida from July 1, 1765 to April 10, 1766. *Transactions of the American Philosophical Society* 33 (1): 1–120.

Basket, James Newton. 1893. Squirrel Hunting. *Atchison Globe* (1 May).

Batty, J.H. 1874. Antelope-Hunting. *The Aldine* 7 (2): 38.

Baumgartner, F.M. 1942. An Analysis of Waterfowl Hunting at Lake Carl Blackwell, Payne County, Oklahoma, for 1940. *Journal of Wildlife Management* 6 (1): 83–91.

Berthel, Mary Wheelhouse. 1935. Hunting in Minnesota in the Seventies. *Minnesota History* 16 (3): 259–271.

Biber, Eric, and Josh Eagle. 2015. When Does Legal Flexibility Work in Environmental Law? *Ecology Law Quarterly* 42 (4): 787–840.

Bissonette, John A., Richard J. Frederickson, and Brian J. Tucker. 1991. American Marten: A Case for Landscape-level Management. In *Wildlife and Habitats in Managed Landscapes*, ed. Jon E. Rodiek and Eric G. Bolen. Washington, DC: Island Press.

Blackwell, Jack. 2018. Policy and Law Relating to Tribal Wildlife Management. In *North American Wildlife Policy and Law*, ed. Bruce D. Leopold, Winifred B. Kessler, and James L. Cummins. Boone and Crockett Club: Missoula.

Boston Journal. 1886. Extraordinary Duck-Hunting. *Daily Evening Bulletin* (10 Mar.), 4.

Branch, E. Douglas. 1929. *The Hunting of the Buffalo*. New York: D. Appleton and Co.

Brasseaux, Carl A., H. Dickson Hoese, and Thomas C. Michot. 2004. Pioneer Amateur Naturalist Louis Judice: Observations on the Fauna, Flora, Geography, and Agriculture of the Bayou Lafourche Region, Louisiana, 1772–1786. *Louisiana History: The Journal of the Louisiana Historical Association* 45 (1): 71–103.

Braverman, Irus. 2015. Conservation and Hunting: Till Death Do They Part? A Legal Ethnography of Deer Management. *Journal of Land Use & Environmental Law* 30 (2): 143–199.

Brook, Barry W., and Corey J.A. Bradshaw. 2006. Strength of Evidence for Density Dependence in Abundance Time Series of 1198 Species. *Ecology* 87: 1445–1451.

Brown, Robert D. 2016. The Politics of Deer-Farming in North Carolina— Lessons Learned. *Wildlife Society Bulletin* 40 (1): 20–24.

———. 2018. The Need for Wildlife Conservation and Policy. In *North American Wildlife Policy and Law*, ed. Bruce D. Leopold, Winifred B. Kessler, and James L. Cummins. Boone and Crockett Club: Missoula.

Bryant, Fred C. 1991. Managed Habitats for Deer in the Woodlands of West Texas. In *Wildlife and Habitats in Managed Landscapes*, ed. Jon E. Rodiek and Eric G. Bolen. Washington, DC: Island Press.

Budiansky, Stephen. 1992. *The Covenant of the Wild: Why Animals Chose Domestication.* New York: William Morrow.

Burroughs, Wilbur G. 1915. Hunting in the Artic and Alaska by E. Marshall Scull. *Bulletin of the American Geographical Society* 47 (2): 139.

Burroughs, R.D., and Laurence Dayton. 1941. Hunting Records for the Prairie Farm, Saginaw County, Michigan, 1937–1939. *Journal of Wildlife Management* 5 (2): 159–167.

Callaghan, Des A., Jeff S. Kirby, and Baz Hughes. 1997. The Effects on Recreational Waterfowl Hunting on Biodiversity. In *Harvesting Wild Species: Implications for Biodiversity Conservation*, ed. Curtis H. Freese. Baltimore: Johns Hopkins University Press.

Campbell, Henry C. 1889a. The Hunting Season. *Yenowine's Illustrated News* (20 Oct.), 1.

———. 1889b. Hunting Noble Game. *Yenowine's Illustrated News* (27 Oct.), 1.

Carlarne, Cinnamon Pinon. 2005. Saving the Whales in the New Millennium: International Institutions, Recent Developments and the Future of International Whaling Policies. *Virginia Environmental Law Journal* 24 (1): 1–48.

Carley, Ira. 1897. Hunting Deer with Dogs. *Milwaukee Daily Sentinel* (11 Feb.), 7.

Carlos, Ann M., and Frank D. Lewis. 1995. Strategic Pricing in the Fur Trade: The Hudson's Bay Company, 1700–1763. In *Wildlife in the Marketplace: The Political Economy Forum*, ed. Terry L. Anderson and Peter J. Hill. New York: Rowman & Littlefield.

Cathlamet Gazette. 1891. Trapping Beaver. *Atchison Daily Champion* (13 Mar.), 7.

Charlotte News. 1894. A Trapping Company. *News and Observer* (13 Mar.).

Chicago Record. 1896. The Trapper's Life. *Galveston Daily News* (18 Aug.), 8.

Chicago Times. 1888. Hunting in the Rockies. *Louisville Courier-Journal* (12 Nov.), 6.

Chico Chronicle. 1886. Fur-Bearing Animals. *Los Angeles Times* (14 Feb.), 2.

Clapham, Phillip J. 2016. Managing Leviathan: Conservation Challenges for the Great Whales in a Post-Whaling World. *Oceanography* 29 (3): 214–225.

Clayton, John. 1694. A Continuation of Mr. John Clayton's Account of Virginia. *Philosophical Transactions* 18: 121–135.

Clemens, Will M. 1886. Hunting Pennsylvania Deer. *Detroit Free Press* (31 July), 6.

Clothier and Furnisher. 1890. Trapping Muskrat. *Atchinson Daily Champion* (10 Aug.), 3.

Cooper, John M. 1929. Canadian Indians Live by Hunting. *Science News-Letter* 16 (448): 286–287.

———. 1939. Is the Algonquian Family Hunting Ground System Pre-Columbian. *American Anthropologist* 41 (1): 66–90.

Cor. New York Herald. 1889. Hunting Wild Turkeys. *Atchison Daily Champion* (6 Jan. 1889), 8.

Crossways, Diana. 1896. Stag Hunting. *Portland Oregonian* (10 May), 15.

Curtis, H.S. 1914. Hunting as Education. *Journal of Education* 80 (10): 260–261.

Davis, John. 1817. *Personal Adventures and Travels Four Years and a Half in the United States of America*. London: W. McDowall.

DeLong, Robert A., and Brian D. Taras. 2009. *Moose Trend Analysis User's Guide*. Alaska Department of Fish and Game.

Denver News. 1883. Riding an Elk. *St. Louis Globe-Democrat* (8 Dec.), 16.

Despain, Don, Douglas Houston, Mary Meagher, and Paul Schullery. 1986. *Wildlife in Transition: Man and Nature on Yellowstone's Northern Range*. Boulder: Roberts Rinehart, Inc.

Detroit Free Press. 1892. A Hunting Episode. *Atchison Globe* (29 Nov.).

Diamond, Joseph E., Thomas Amorosi, and David Perry. 2016. Late Woodland Subsistence at the Wolfersteig Site: A Multi-Component Site on the Esopus Creek. *Archaeology of Eastern North America* 44 (1): 131–160.

Dickinson, Nate. 1993. *Common Sense Wildlife Management: Discourses on Personal Experiences*. Altamont, NY: Settle Hill Publishing.

Disagreeable Experience. 1892. *Portland Oregonian* (28 Nov.), 6.

Dudley, Paul, and John Chamberlayne. 1720–1721. A Description of the Moose-Deer in America. *Philosophical Transactions* 31: 165–168.

Dunaway, Wilma A. 1994. The Southern Fur Trade and the Incorporation of Southern Appalachia into the World-Economy, 1690–1763. *Fernand Braudel Center Review* 17 (2): 215–242.

Dunraven, Lord. 1879. Moose and Cariboo Hunting in Colorado and Canada. *Journal of the American Geographical Society of New York* 11: 334–368.

———. 1881. Hunting the Moose. *Daily Central City Register* (16 March).

Eddy, J.W. 1924. *Hunting on Kenai Peninsula*. Seattle: Lowman & Hanford Co.

Edwards, George. 1753–1754. A Letter to Mr. Peter Collinson, F.R.S. Concerning the Pheasant of Pensylvania [sic], and the Otis Minor. *Philosophical Transactions* 48: 499–503.

Erie Dispatch. 1884. Hunting in Pennsylvania. *Cleveland Daily Herald* (13 Jan.), 12.

Exchange. 1888. Sylvester Scott, Bear Hunter. *Milwaukee Journal* (16 March).

F.E.S. 1878. Hunting Experience. *Daily Evening Bulletin* (27 Feb.).

Ferril, Will C. 1893. The Hunting Season. *Rocky Mountain News* (1 Oct.), 9.

Fiedel, Stuart J. 2001. What Happened in the Early Woodland? *Archaeology of Eastern North America* 29 (1): 101–142.

Forest and Stream. 1882. The Trapper's Last Shot. *Daily Inter Ocean* (1 May), 10.

———. 1889. The Bear-Hunting of Today. *Rocky Mountain News* (29 Nov.), 2.

———. 1892. Hunting with a Camera. *Daily Inter Ocean* (14 May), 2.

———. 1895. Trappers as Packers. *Milwaukee Daily Sentinel* (15 Dec.), 13.

————. 1896. The Mastodon Not Extinct. *Milwaukee Journal* (27 Nov.).

Fort Collins Express. 1881. One Season's Hunting. *St. Louis Globe-Democrat* (2 Jan. 1881), 11.

Foster, H. Thomas, III, and Arthur D. Cohen. 2007. Palynological Evidence of the Effects of the Deerskin Trade on Forest Fires During the Eighteenth Century in Southeastern North America. *American Antiquity* 72 (1): 35–51.

Fryxell, John M., David J.T. Hussell, Anne B. Lambert, and Peter C. Smith. 1991. Time Lags and Population Fluctuations in White-Tailed Deer. *Journal of Wildlife Management* 55 (3): 377–385.

Gainesville Florida Eagle. 1880. Trapping Beavers in North Florida. *Daily Evening Bulletin* (7 Apr.).

Gallardo, Julio C. 2018. Jurisdictions in Mexico. In *North American Wildlife Policy and Law*, ed. Bruce D. Leopold, Winifred B. Kessler, and James L. Cummins. Boone and Crockett Club: Missoula.

Gallman, Robert E., and Paul W. Rhode. 2019. *Capital in the Nineteenth Century*. Chicago: University of Chicago Press.

Garshells, David L., and Hank Hristienko. 2006. State and Provincial Estimates of American Black Bear Numbers Versus Assessments of Population Trend. *Ursus* 17 (1): 1–7.

Gillespie, Alexander. 2005. *Whaling Diplomacy: Defining Issues in International Law*. New York: Edward Elgar.

Hairr, John. 2011. John Lawson's Observations on the Animals of Carolina. *North Carolina Historical Review* 88 (3): 312–332.

Harper's Magazine. 1870. Hunting the Canvas-Back. *Lowell Daily Citizen* (3 Feb.).

Hodak, Marc, and Jack Masterson. 2021. *The Seeds of Their Own Destruction: Lessons from Utopian Experiments in Nineteenth-Century America, Part 1—The Other American Dream*. SSRN Working Paper.

Holmes, Thomas. 1893. Hunting the Polecat. *Scientific American* 68 (14): 218.

Holyoke, John. 2019. Maine Hunters on Track to Harvest More than 30,000 Deer This Year. *Bangor Daily News* (21 Nov.).

Hough, E. 1889. Deer Coursing with Greyhounds. *Outing: An Illustrated Monthly Magazine of Sport, Travel and Recreation* 14: 426–435.

Huffman, James L. 1995. In the Interests of Wildlife: Overcoming the Tradition of Public Rights. In *Wildlife in the Marketplace: The Political Economy Forum*, ed. Terry L. Anderson and Peter J. Hill. New York: Rowman & Littlefield.

Hunt with the Yankton Sioux. 1873. Indian Hunting. *Lowell Daily Citizen* (22 Sept.).

Imperio, Simona, Massimiliano Ferrante, Alessandra Grignetti, Giacomo Santini, and Stefano Focardi. 2010. Investigating Population Dynamics in Ungulates: Do Hunting Statistics Make Up a Good Index of Population Abundance? *Wildlife Biology* 16: 205–214.

Indianapolis Journal. 1886. Trappers' Earnings Fifty Years Ago. *St. Louis Globe-Democrat* (4 Nov.), 10.

Isenberg, Andrew C. 2000. *The Destruction of the Bison, 1750–1920.* New York: Cambridge University Press.

Johnsen, D. Bruce. 2009. Salmon, Science, and Reciprocity on the Northwest Coast. *Journal of Ecology and Society* 14 (43). http://www.ecologyandsociety.org/vol14/iss2/art43.

———. 2016. Reciprocity and Fractional Reserve Banking Among Northwest Coast Tribes. In *Unlocking the Wealth of Indian Nations*, ed. Terry Anderson. New York: Lexington.

Johnston, John, J.W. Ormsby, H.N. Campbell, Edward Silverman, H.T. Drake, Stephen Meunier, C.H. Mathews, Alfred James, and R.G. Richter. 1896. The Hunting of Deer. *Milwaukee Journal* (7 Nov.).

Jones, Hugh. 1699. Part of a Letter from the Reverend Mr. Hugh Jones to the Reverend Dr. Benjamin Woodroofe, F.R.S. Concerning Several Observables in Maryland. *Philosophical Transactions* 21: 436–442.

Jonker, Sandra A., Robert M. Muth, John F. Organ, Rodney R. Zwick, and William F. Siemer. 2006. Experiences with Beaver Damage and Attitudes of Massachusetts Residents Toward Beaver. *Wildlife Society Bulletin* 34 (4): 1009–1021.

Kansas City Star. 1890. Hunting Wild Ponies. *Bismarck Tribune* (19 Nov.), 2.

Kay, Charles E. 1997. Aboriginal Overkill and the Biogeography of Moose in Western North America. *Alces* 33: 141–164.

Kellogg, Vernon. 1926. Hunting Bighorn with a Camera. *Scientific Monthly* 23 (2): 112–116.

King, J.S. 1899. Lively Experience with an Old Hunter with Bears. *Denver Evening Post* (7 May), 16.

Kinietz, Vernon. 1940. Notes on the Algonquian Family Hunting Ground System. *American Anthropologist* 42 (1): 179.

Kluender, Richard A., Philip A. Tappe, and Michael E. Cartwright. 1992. Long-Term White-Tailed Deer Harvest Trends for the Southcentral United States. *Journal of the Arkansas Academy of Science* 46 (5): 49–52.

Knapp, Charles. 1935. Hogs Roman and Modern Boar Hunting, Ancient and Modern. *Classical Weekly* 28 (11): 81–84.

Koch, Paul L., and Anthony D. Barnosky. 2006. Late Quaternary Extinctions: State of the Debate. *Annual Review of Ecology, Evolution, and Systematics* 37: 215–250.

Kreppel, Peter. 1897. About Duck Hunting. *Milwaukee Journal* (17 Feb.), 8.

Laliberte, Andrea S., and William J. Ripple. 2003. Wildlife Encounters by Lewis and Clark: A Spatial Analysis of Interactions Between Native Americans and Wildlife. *BioScience* 53 (10): 994–1003.

————. 2004. Range Contractions of North American Carnivores and Ungulates. *BioScience* 54 (2): 123–138.

Leopold, Bruce D. 2018. History of Wildlife Policy and Law Through Colonial Times. In *North American Wildlife Policy and Law*, ed. Bruce D. Leopold, Winifred B. Kessler, and James L. Cummins. Boone and Crockett Club: Missoula.

Leopold, Bruce D., Winifred B. Kessler, and James L. Cummins. 2018. Preface. In *North American Wildlife Policy and Law*, ed. Bruce D. Leopold, Winifred B. Kessler, and James L. Cummins. Boone and Crockett Club: Missoula.

Lewiston Journal. 1881. Maine's Fur Industry. *St. Louis Globe-Democrat* (24 Dec.), 7.

————. 1885. A Bear Hunter. *Bismarck Tribune* (24 Oct.).

————. 1886. A Famous Maine Hunter. *St. Louis Globe-Democrat* (17 July), 4.

London Saturday Review. 1887. American Moose Hunting. *St. Louis Globe-Democrat* (14 Sept.), 6.

Lueck, Dean L. 1995. The Economic Organization of Wildlife Institutions. In *Wildlife in the Marketplace: The Political Economy Forum*, ed. Terry L. Anderson and Peter J. Hill. New York: Rowman & Littlefield.

Mack, Julie. 2017. Michigan Ranks No. 2 in 2016 Deer Harvest, and Other Deer-Hunting Facts. *Michigan Live* (6 Nov.).

MacLeod, William Christie. 1922. The Family Hunting Territory and Lenape Political Organization. *Anthropologist* 24 (4): 448–463.

Madrigal, T. Cregg, and Julie Zimmermann Holt. 2002. White-Tailed Deer Meat and Marrow Return Rates and Their Application to Eastern Woodlands Archaeology. *American Antiquity* 67 (4): 745–759.

Malley, Edward. 1895. Delightful Hunting Trip. *Boston Daily Advertiser* (17 Oct.), 7.

Mancall, Peter C. 2013. The Raw and the Cold: Five English Sailors in Sixteenth-Century Nanavut. *William and Mary Quarterly* 70 (1): 3–40.

Mancall, Peter C., and Thomas Weiss. 1999. Was Economic Growth Likely in Colonial British America? *Journal of Economic History* 59 (1): 17–40.

McCarty, George S. 1934. Scientific Methods of Game Breeding Will Make Good Hunting. *Scientific American* 151 (5): 234–235.

McHugh, J.L. 1977. Rise and Fall of World Whaling: The Tragedy of the Commons Illustrated. *Journal of International Affairs* 31 (1): 23–33.

Memphis Avalanche. 1883. Over Five Thousand Pigeons Killed. *St. Louis Globe-Democrat* (15 Dec.), 16.

Michelson, Truman. 1921. Note on the Hunting Territories of the Sauk and Fox. *American Anthropologist* 23 (2): 238–239.

Mighels, Philip V. 1897. Hints for Young Trappers. *Salt Lake Semi-Weekly Tribune* (11 May), 13.

Miles, Nelson A. 1895. Hunting Large Game. *North American Review* 161 (467): 484–492.

Millan, J. 1744. *The Present State of the Country and Inhabitants, Europeans and Indians, of Louisiana*. London: J. Millan.

Miller, Robert, Jr. 2012. *Reservation 'Capitalism': Economic Development in Indian Country*. New York: Praeger.

Moore, J.R. 1897. A Deer Hunting Law. *Milwaukee Daily Sentinel* (5 Feb.), 4.

Mulderink, Earl F. 2012. *New Bedford's Civil War*. New York: Fordham University Press.

Munkittrick, R.K. 1893. A Champion Hunter. *Yenowine's Illustrated News* (30 Dec.), 6.

Murdoch, W.G. Burn. 1917. *Modern Whaling and Bear Hunting*. London: Seeley, Service & Co.

Nagaoka, Lisa, Torben Rick, and Steve Wolverton. 2018. The Overkill Model and Its Impact on Research. *Ecology and Evolution* 8 (19): 9683–9696.

Natchez Free Trader. 1852. *Missouri Courier* (4 March).

Nesbit, William. 1926. *How to Hunt with the Camera*. London: G. Allen & Unwin Limited.

Neumann, Thomas W. 1985. Human-Wildlife Competition and the Passenger Pigeon: Population Growth from System Destabilization. *Human Ecology* 13 (4): 389–410.

New Orleans Time Democrat. 1883. Millions of Mallards. *St. Louis Globe-Democrat* (15 Dec.), 16.

New York Commercial. 1841. Venison. *North American* (27 Nov.).

———. 1875. Fox-Hunting. *St. Louis Globe-Democrat* (14 Nov.), 12.

New York Commercial Advertiser. 1860. The Buffalo Robe Trade. *Milwaukee Daily Sentinel* (24 Aug.).

New York Mail. 1887. Owl Hunting. *Boston Daily Advertiser* (25 Nov.), 5.

New York Mail and Express. 1886. Hunting the Gray Squirrel. *St. Louis Globe-Democrat* (9 Dec.), 5.

New York Sun. 1878. Buffalo Hunting. *St. Louis Globe-Democrat* (24 Feb.), 12.

———. 1884. Hunting Notes. *St. Louis Globe-Democrat* (11 Aug.), 7.

———. 1889. Trapping Turkeys. *Atchison Daily Champion* (23 Jan.), 3.

———. 1897. New Hunting Rifles. *Milwaukee Journal* (5 Nov.), 7.

New York Times. 1877. Hunting in Florida. *St. Louis Globe-Democrat* (14 Jan.), 9.

———. 1884. The Hunters of Maine. *St. Louis Globe-Democrat* (10 May), 12.

———. 1885. An Apache Hunter. *Bismarck Tribune* (28 Apr.).

———. 1887. Trapping the Grizzly. *Daily Inter Ocean* (10 Nov.), 24.

New York Tribune. 1896. The Hunting Season. *Daily Inter Ocean* (22 Nov.), 27.

Nye, Bill. 1887. Fox Hunting. *Rocky Mountain News* (23 Oct.), 9.

O. 1889. Deer Hunting. *Portland Oregonian* (13 Jan.).

Organ, John F. 2018. The North American Model of Wildlife Conservation. In *North American Wildlife Policy and Law*, ed. Bruce D. Leopold, Winifred B. Kessler, and James L. Cummins. Boone and Crockett Club: Missoula.

Organ, John F., et al. 2012. The North American Model of Wildlife Conservation. *The Wildlife Society Technical Review* 12-04.

Ottaway, Andy. 2013. *Commercial Whaling. The Global Guide to Animal Protection*. Urbana: University of Illinois Press.

Oxley, J. Macdonald. 1888. Hunting the Moose. *Bismarck Tribune* (20 March).

Palmer, E. Laurence. 1939. Farm Forest Facts. *Cornell Rural School Leaflet* 33 (2): 1–32.

Pavao-Zuckerman, Barnet. 2007. Deerskins and Domesticates: Creek Subsistence and Economic Strategies in the Historic Period. *American Antiquity* 72 (1): 5–33.

Pellet, Frank. 1938. *History of American Beekeeping*. Ames: Iowa Collegiate Press.

Perttula, Timothy K., Cathy J. Crane, and James E. Bruseth. 1982. A Consideration of Caddoan Subsistence. *Southeastern Archaeology* 1 (2): 89–102.

Pluckhahn, Thomas J., J. Matthew Compton, and Mary Theresa Bonhage-Freund. 2006. Evidence of Small-Scale Feasting from the Woodland Period Site of Kolomoki, Georgia. *Journal of Field Archaeology* 31 (3): 263–284.

Portland Oregonian. 1883. About Bears. *St. Louis Globe-Democrat* (10 Nov.), 4.

Pressly, Paul M. 2013. *On the Rim of the Caribbean: Colonial Georgia and the British Atlantic World*. Athens: University of Georgia Press.

Reagan, Albert B. 1919–1921. Hunting and Fishing of Various Tribes of Indians. *Transactions of the Kansas Academy of Science* 30: 443–448.

Reeves, Randall D., and Tim D. Smith. 2006. *Whales, Whaling, and Ocean Ecosystems*. Berkeley: University of California Press.

Reitz, Elizabeth J., and Gregory A. Waselkov. 2015. Vertebrate Use at Early Colonies on the Southeastern Coasts of Eastern North America. *International Journal of Historical Archaeology* 19 (1): 21–45.

Richards, John F. 2014. *The World Hunt: An Environmental History of the Commodification of Animals*. Berkeley: University of California Press.

Robinson, H.M. 1879. Indian Trappers, Hudson Bay. *Galveston Daily News* (28 May).

Rockwell, Robert H. 1922. Hunting the Big Brown Bear. *Brooklyn Museum Quarterly* 9 (1): 1–23.

———. 1923. Sheep Hunting in Alaska. *Brooklyn Museum Quarterly* 10 (2): 71–82.

———. 1924. Moose Hunting in Alaska. *Brooklyn Museum Quarterly* 11 (2): 76–81.

Roosevelt, Theodore. 1892. Antelope Hunting. *Daily Inter Ocean* (24 Jan.), 27.

Ross, W. Gillies. 1979. The Annual Catch of Greenland (Bowhead) Whales in Waters North of Canada, 1719–1915. *Arctic* 32 (2): 91–121.

Rucker, Randall, Walter N. Thurman, and Michael Burgett. 2019. Colony Collapse and the Consequences of Bee Disease: Market Adaptation to Environmental

Change. *Journal of the Association of Environmental and Resource Economists* 6 (5): 927–960.

S. H. 1833. Bull-Hunting in Washitaw. *Maryland Gazette* (18 July).

San Francisco Call. 1885. Traps and Trappers. *Galveston Daily News* (5 Nov.), 6.

San Francisco Examiner. 1887. Trapping Black Bear. *St. Louis Globe-Democrat* (20 July), 6.

———. 1888. Hunting in Idaho. *Rocky Mountain News* (21 March), 3.

Savage, James. 1825. *The History of New England from 1630 to 1649 by John Winthrop*. Boston: Phelps and Farnham.

Savage, Henry L. 1933. Hunting in the Middle Ages. *Speculum* 8 (1): 30–41.

Scanland, J.M. 1893. Hunting the Elk. *Bismarck Tribune* (23 July), 4.

Shields, G. O. 1887. Hunting the Grizzly. *St. Louis Daily Globe-Democrat* (31 Jul.), 28.

Shoemaker, Nancy. 2005. Whale Meat in American History. *Environmental History* 10 (2): 269–294.

Smalley, Andrea L. 2016. 'They Steal Our Deer and Land': Contested Hunting Grounds in the Trans-Appalachian West. *Register of the Kentucky Historical Society* 114 (3/4): 303–339.

Smith, De Cost. 1889. Onondaga Superstitions. Hunting. *Journal of American Folklore* 2 (7): 282–283.

Smith, Paul A. 2019. Deer Population at Potential Record High as Hunters Set Sights on 2019 Wisconsin Gun Season. *Milwaukee Sentinel Journal* (16 Nov.).

Sokos, Christos K., M. Nils Peterson, Periklis K. Birtsas, and Nikolas D. Hasanagas. 2014. Insights for Contemporary Hunting from Ancient Hellenic Culture. *Wildlife Society Bulletin* 38 (3): 451–457.

Southwell, Robert. 1686–1692. The Method the Indians in Virginia and Carolina Use to Dress Buck and Doe Skins. *Philosophical Transactions* 16: 532–533.

Sowell, Thomas. 2009. *Intellectuals and Society*. New York: Perseus Group.

Special Correspondence. 1887. Deer Hunting. *St. Louis Globe-Democrat* (7 Aug.), 20.

Speck, Frank G. 1915. The Family Hunting Band as the Basis of Algonkian Social Organization. *American Anthropologist* 17 (2): 289–305.

———. 1923. Mistassini Hunting Territories in the Labrador Peninsula. *American Anthropologist* 25 (4): 452–471.

Speck, Frank G., and Loren C. Eiseley. 1939. Significance of Hunting Territory Systems of the Algonkian in Social Theory. *American Anthropologist* 41 (2): 269–280.

———. 1942. Montagnais-Naskapi Bands and Family Hunting Districts of the Central and Southeastern Labrador Peninsula. *Proceedings of the American Philosophical Society* 85 (2): 215–242.

Spiess, Arthur, Kristin Sobolik, Diana Crader, John Mosher, and Deborah Wilson. 2006. Cod, Clams and Deer: The Food Remains from Indiantown Island. *Archaeology of Eastern North America* 34 (1): 1471–1487.

Sporting Journal. 1885. Hunting Wild Turkeys. *Los Angeles Times* (31 May).

St. Louis Globe-Democrat. 1894. Hunting Wild with Tame Turkeys. *Bismarck Tribune* (26 Feb.), 2.

St. Paul Pioneer Press. 1893. The Successful Hunter. *Bangor Daily Whig and Courier* (8 Dec.).

Staunton Vindicator. 1877. A Great Hunter Killed. *St. Louis Globe-Democrat* (17 Dec.), 2.

Stothers, David M., and Timothy J. Abel. 1993. Archaeological Reflections of the Late Archaic and Early Woodland Time Periods in the Western Lake Erie Region. *Archaeology of Eastern North America* 21 (1): 25–109.

Swanson, Drew A. 2018. *Beyond the Mountains: Commodifying Appalachian Environments*. Athens: University of Georgia Press.

Taylor, H.W. 1888. Turkey Hunting. *Milwaukee Daily Sentinel* (1 Jan.).

Thomas, William S. 1906. *Hunting Big Game with Gun and with Kodak: A Record of Personal Experiences in the United States, Canada, and Mexico*. New York: G.P. Putnam's Sons.

Trapper's Guide. 1865. *Bangor Daily Whig and Courier* (30 Sept.).

Trefethen, James B. 1975. *An American Crusade for Wildlife*. New York: Winchester Press.

Usner, Daniel H., Jr. 1985. The Deerskin Trade in French Louisiana. *Proceedings of the Meeting of the French Colonial Historical Society* 10: 75–93.

———. 1992. *Indians, Settlers, and Slaves in a Frontier Exchange Economy: The Lower Mississippi Valley Before 1783*. Chapel Hill: University of North Carolina Press.

Vicksburg Commercial Herald. 1892. Hunting the Hunter. *New Orleans Daily Picayune* (7 Nov. 1892), 7.

W. S. 1915. McIlhenny's The Wild Turkey and Its Hunting. *The Auk* 32 (1): 115–116.

———. 1924. *Recollections of Fifty Years Hunting and Shooting* by Wm. B. Mershon. *The Auk* 41 (2): 365.

Wales, William. 1770. Journal of a Voyage, Made by Order of the Royal Society, to Churchill River, on the North-west Coast of Hudson's Bay. *Philosophical Transactions* 60: 100–136.

Walsh, Virginia M. 1999. Illegal Whaling for Humpbacks by the Soviet Union in the Antarctic, 1947–1972. *Journal of Environment & Development* 8 (3): 307–327.

Waselkov, Gregory A. 1978. Evolution of Deer Hunting in the Eastern Woodlands. *Midcontinental Journal of Archaeology* 3 (1): 15–34.

Webb, G. Kent. 2018. Searching the Internet to Estimate Deer Population Trends in the U.S., California, and Connecticut. *Issues in Information Systems* 19 (2): 163–173.

West Bloomfield. 1889. An Adirondack Hunter. *Daily Inter Ocean* (22 Dec.)

Western Hunter. 1889. Hunting Buffaloes. *Milwaukee Sentinel* (30 Aug.), 4.

Weyawega Times. 1872. An Old Trapper's Adventure. *Lowell Daily Citizen* (2 Nov.).

Wheat, Joe Ben, Harold E. Malde, and Estella B. Leopold. 1972. The Olsen-Chubbuck Site: A Paleo-Indian Bison Kill. *Memoirs of the Society for American Archaeology* 26: 1–180.

Wheeler, E.P. 1930. Journeys About Nain Winter Hunting with the Labrador Eskimo. *Geographical Review* 20 (3): 454–468.

Whig Man. 1899. Hunting Season. *Bangor Daily Whig and Courier* (9 May), 2.

White, Richard. 1983. *The Roots of Dependency: Subsistence, Environment, and Social Change Among the Choctaws, Pawnees, and Navajos*. Lincoln: University of Nebraska Press.

Whitney, Leon F. 1931. The Raccoon and Its Hunting. *Journal of Mammalogy* 12 (1): 29–38.

Williams, M.C. 1889. Happy Hunting Grounds. *Daily Inter Ocean* (1 Sept.), 21.

Williams, A., and D. Bugbee. 1865. The Trapper's Guide. *Bangor Daily Whig and Courier* (19 Sept.).

Winnebago County Press. 1870. The Muskrat Catch. *Milwaukee Daily Sentinel* (10 May).

Winona Republican. 1873. The Pigeon Business. *Milwaukee Daily Sentinel* (23 May), 2.

Wright, Charles. 1868a. Bears and Bear-Hunting. *The American Naturalist* 2 (3): 121–124.

———. 1868b. Deer and Deer-Hunting in Texas. *The American Naturalist* 2 (9): 466–476.

Wright, Robert E. 2019. America's Fur Business, Parts I, II, and III. *Fur Traders & Rendezvous.* https://www.alfredjacobmiller.com/explore/americasfurbusiness1/.

Young, Kimball, and Thomas D. Cutsforth. 1928. Hunting Superstitions in the Cow Creek Region of Southern Oregon. *Journal of American Folklore* 41 (160): 283–285.

Zink, Robert M. 2014. *The Three-Minute Outdoorsman: Wild Science from Magnetic Deer to Mumbling Carp*. St. Paul: University of Minnesota Press.

CHAPTER 4

The Dangers of Democracy

Abstract Chapter 4: Warns that the facile application of "democracy" to wildlife management poses grave risks for wildlife, especially wild game species, by threatening the authority of scientific wildlife management and further eroding the wildlife dollars volunteered by consumptive users.

Keywords North American Wildlife Conservation Mode • Democracy • Scientific wildlife management • Voluntary contributions • Consumptive use • Political economy of conservation

As political economist Richard E. Wagner put it, "democracy may serve to secure liberty and property, but it may also serve to undermine them. … If individual rights to resources exist, it is only because the government has chosen to tolerate such rights" (Wagner 1998, 316, 319). As Americans and Canadians learned in 2020, governments stand ready to deny them longstanding rights for flimsy reasons backed by little more than pseudo-science (Earle 2020).

So it is hardly a stretch to note that the decline of public support for trapping (Vantassel et al. 2010, 938) may portend public animosity toward hunting and fishing that could eventually lead to curtailed or banned consumptive use rights (Huffman 1995, 29). A key problem is the changing conception of one of the key sisters of the NAWCM, "democracy." What

© The Author(s), under exclusive license to Springer Nature Switzerland AG 2022
R. E. Wright, *The History and Evolution of the North American Wildlife Conservation Model*,
https://doi.org/10.1007/978-3-031-06163-9_4

has made the NAWCM work, especially in the United States, is a "user-pay, user-benefit model," not allowing non-payers (who are often non-users with limited understanding of the issues involved) to dictate policy, even if they constitute a majority of voters (Organ et al. 2012, 9).

Originally, "democracy" meant that consumptive users would maintain a voice in wildlife managers' decision-making. More recently, however, some commentators have conflated "democracy" with equal economic access when deriding fee-based hunting (Organ et al. 2016, 11–12). They seem not to understand that farmers have leased land to hunters by the season or day since at least the Great Depression (Palmer 1939, 17), when institutions called Cooperative Hunting Exchanges, apparently kin to nineteenth-century hunting clubs that paid farmers in a share of the game harvested (Anon. 1894), also operated (Burroughs 1937, 21). Fishing and hunting clubs also outright purchased or cash-leased prime lands and fishing holes, some near urban centers and some in the wilds of the West or Canada, for the exclusive use of their members (Anon. 1876, 1895a; Williams 1889). Members of the country's hunting clubs, which numbered in the hundreds by the 1890s, gained access to tens of thousands of acres of land, lodgings, and comradery for a month or two each year. Tellingly, it was said that the amount paid for membership was more about the quality of the members of the club than the quality of the hunting experience (Anon. 1895b).

Other recent commentators call for "fair" or "equitable" governance of wildlife, meaning a form of "majority rule" in which all citizens, not just consumptive users, get an equal vote (Decker et al. 2017, 823). It is well understood, however, that majority rule can distort group preferences because votes are equally weighted (Huffman 1995, 28). This is an especially important point where the preferences of one of the major stakeholders, wildlife, cannot effectively be discerned but rather must be assumed to be the maximization of the resources expended on their behalf. It is not that wild creatures need money per se but that the monetary values placed upon them indicate their worth to various human stakeholders.

Imagine, for example, a scenario where 1000 people are willing to pay $1 on average to ban the hunting of species X or all consumptive use in area Y, but 100 people are willing to spend $1000 each on average for the right to engage in the scientifically managed harvest of X or to continue consumptive use in area Y. In a straight vote, the anti-hunters would win 1000 to 100 but wildlife would then receive only $1000 while in the latter case it would receive $100,000 of support. Clearly, the latter provides more material support for wildlife. (If you think otherwise, see Budiansky 1992: 127–52 and Dickinson 1993, 54). A major problem with wildlife management in Canada is that it does not earmark excise taxes as the United States does, leaving wildlife to compete with humans for funding. "The result," one recent study concluded, "is that wildlife does not, in almost all circumstances, receive what its proponents and managers believe is its due" (Organ et al. 2012, 7).

Decision-making based on actual resource allocation, rather than mere virtue signaling, will be of crucial importance to wildlife if consumptive users continue to decline as a proportion of the total population (Decker et al. 2017, 822). It is one thing for a person to assert that they "value" something, like wild places or things, and quite another for them to actually transfer resources to it, especially when they know little about it, or how it functions (Organ 2018, 131). Only if people truly value the wild more than the domestic can wildlife be saved from destruction, a lesson that is crystal clear on the hardscrabble margins of the global dry lands, especially in poor places like southern Africa (Child 2019, 6).

Voters with a limited stake in outcomes may also injure people. For example, voters in Massachusetts effectively banned beaver trapping for non-scientific reasons. The costs of that ban, more beaver dams that destroyed private property, were borne solely by affected homeowners and may well lead to poaching, which by definition is illegal and difficult to monitor, instead of the scientifically managed culling that had been succeeding before the tyranny of the majority was allowed to dictate policy (Jonker et al. 2006, 1019). Such unchecked democracy, or majority rule, can imperil the foundations of American society (Holcombe 2019).

Bans on hunting based on verifiable safety concerns, for example, are no doubt just, but what prevents the imposition of unreasonable safety restrictions to ban hunting indirectly? Who gets to decide when safety restrictions are justified and when they are not? What about deciding between the legitimate safety concerns of homeowners versus motorists (Braverman 2015, 160–61)?

Putting aside bigger questions of the long-term viability of democratic governance (Wagner 1998), voter preferences often reduce to a struggle over resources couched in rhetorical garb. What native Alaskans seek, for example, is not to "have a meaningful role" in wildlife management (Brooks and Bartley 2016, 517), they want to either exclude other consumptive users from their territories or at least get a larger share of their license fees, with the goal of ensuring that wildlife resources will not be overharvested. Locals of course suffer most from unsustainable practices and also usually have learned local game management best practices (Ostrom 1990; Child 2019, 7). When analyzed economically, instead of emotionally, rational solutions to resource conflicts are more likely to emerge.

Wildlife manager Steve Joule recently noted that "deer management is not that complicated; it's the people management that's extremely complicated." Hunters want more deer but farmers want fewer. A non-hunting non-farmer who lives nearby "wants to see deer but doesn't want them getting too close," while an animal rights group "wants deer just to be left alone completely," no matter the costs (see, e.g., Czech 2000), while yet "another group … thinks you should reintroduce wolves to maintain the population." Meanwhile, motorists are dying "and the municipality … doesn't want to do anything with the park because the park is for walking your dog" (Braverman 2015, 189). That is all true, but Joule could have added that many wildlife enthusiasts also want to divert the dollars of consumptive users (fishers, hunters, trappers) to the maintenance of non-game species.

NAWCM critics complain that it privileges consumptive over non-consumptive users like bird watchers (Braverman 2015, 145; Feldpausch-Parker et al. 2017). They correctly point out that the NAWCM does not easily apply to non-game species, the populations of which are not subject to human predation much less the tragedy of the commons. The fate of such species is often tied to conserving or reconstructing natural habitat, which may require extending commons left, or returned, to their natural state, including the re-introduction of natural fire regimes and bison and other native grazers (Jahn 1991, xviii; Harris 1984; Brennan and Kuvlesky 2005, 7–8).

Consumptive user dollars conserve habitat through license fees, equipment taxes, and voluntary donations to sundry Species X Forever NGOs and other wildlife nonprofits (Cummins 2018) but that means that the habitat needs of game species, which may differ in important respects from that of non-game species, is prioritized (Dickinson 1993, 23, 77–78). Similarly, rural landowners who hunt voluntarily create significant habitat, though mostly focused on immediate hunting goals, like food plots, rather than longer-term management, like prescribed burns and weed control (Golden et al. 2013).

Some species of non-game grassland birds need much larger patches of contiguous habitat than pheasants, quail, and other upland game birds require and hence have seen their habitats shrink as ditch-to-ditch agriculture spreads across the plains, range deterioration increases in the arid West, and afforestation occurs in the East (Brennan and Kuvlesky 2005, 4–5). Moreover, hunting and other forms of pest control of prairie dogs injures the habitat of many non-game upland bird species (Brennan and Kuvlesky 2005, 5). Such conflicts exacerbate tensions between important stakeholders, including bird watchers, hunters, and landowners (Brennan and Kuvlesky 2005, 9), conflicts that hunters often win because their funding has captured their regulators (WMs), many of whom are hunters themselves (Braverman 2015, 145).

Where critics (Feldpausch-Parker et al. 2017) see "ideology" and "power," however, others see voluntary association. Where they see mostly non-hunters (Decker et al. 2017, 822), others see people who invest large sums in wildlife conservation not as an abstract ideal but to further their own consumption. What the critics want is to divert even more of those voluntary hunter dollars, which are still the primary source of conservation funds (Schorr et al. 2014, 944–45; Braverman 2015, 151–52), to the conservation of non-game animals.

Diverting hunter dollars to non-hunting uses, however, threatens both game and non-game species by inducing hunters to quit the sport by changing their cost-benefit analysis. Hunters and other consumptive users provide most wildlife management dollars. They, like other consumers, are sensitive to prices, like licensing fees (Schorr et al. 2014, 947, 950), but also to a range of other costs as well as perceived benefits. Hunters might be perfectly willing to pay, say, $500 for a cow elk tag with a 50 percent probability of success only if they believe that if their tags go unfilled their money will improve the elk population. If they think unfilled tags profit Greenpeace or other free riders, like birders who do not buy duck stamps (Lueck 1995, 4; Vrtiska et al. 2013, 386), they will need a lower license fee or a higher probability of success in order to be induced to volunteer their dollars.

Voluntarily funding a public good like wildlife management is not hegemonic, as critics following Foucault and kindred spirits claim (Feldpausch-Parker et al. 2017, 36–37), it is a form of democracy by the dollar. Consumptive users "vote" with their licensing and excise fee dollars (Pittman-Robertson Act of 1937 and the Dingell-Johnson Act of 1950 [Lueck 1995, 4]) to support wild game consumption (Winkler and Warnke 2013, 461). Those interested in preserving non-game species will find consumptive users no barrier, so long as the preservationists raise

their own funds to the extent that their projects do not aid in the propaga-
tion of game species. Some consumptive users may also donate money to
non-consumptive causes, but only as a separate voluntary act. Attempts to
use the political process to divert dollars donated to consumptive uses for
non-consumptive purposes, or to groups with a known anti-consumptive
agenda, have met with resistance and, when that failed, the withdrawal of
voluntary dollars from the wildlife conservation system through increased
poaching or decreased recruitment of new consumptive users and
decreased retention of existing ones.

Some non-consumptive users would also like to use majoritarian
democracy to block private wildlife management innovation, prompting
warnings, for example, that captive cervid facilities might make voters
more likely to consider deer domesticated animals instead of wildlife
(Organ et al. 2016, 11). That seems unlikely because wild versions of species that have been domesticated for decades or even centuries are still
considered huntable quarry where they roam free. Alligators, beef cattle,
goats, hogs, rabbits, sheep, and turkeys come immediately to mind.
Moreover, captive cervids, although long acknowledged to possess a low
degree of "domesticity" (Wright 1868, 466–67), are not new. Eighteenth-
century South Carolina newspapers frequently mentioned the loss or pur-
chase of "tame deer" (Dunbar 1962) and an Oregon farmer in the 1880s
noted that fawns were "often" captured and closely studied (O. 1889). In
1895, a newspaper reported that "fawn trappers" were in trouble, not
because they had captured young deer but because they did it on an Indian
Reservation (Hennessey). That same year, somebody suggested that deer
parks be established in wooded areas around cities that could not otherwise be developed without endangering the quality of the urban water
supply for the explicit purpose of supplying markets with venison (Anon.
1895c). The following year, park rangers in Baltimore captured three deer
in a feed pen trap for translocation to a park in Trenton, New Jersey that
wanted two does and a buck (Baltimore Sun 1896). That same year
(1896), the *Helena Independent* explained how ranchers who wanted private herds of wapiti hired skiers to use lariats to capture them alive when
winter snow conditions were optimal.

Species with reproduction assisted by humans, like pheasants and trout,
are still considered wild game too, even if some may consider stocked
game less valuable than those wild born (Dickinson 1993, 26, 54; Knox
2011, 48; Peterson et al. Peterson et al. 2016, 431). Interestingly, not all

of the 5500 plus captive cervid facilities active in the United States use artificial methods of insemination (Adams et al. 2016, 15).

Bears were captured young by killing their mothers (Erie Dispatch 1884) but many bear captures, unlike cervid captures, were motivated by the menagerie (private zoo, a la the Tiger King) trade (New York Sun 1888). Menageries and public zoos raise a host of economic and ethical issues not addressed here, but nobody says that lions, tigers, and bears are not wild creatures because Joe Exotic and the Washington Zoo keep representatives of the species in cages. Over time, though, artificial selection will tend to domesticate any species bred for human purposes, specifically toward less aggressive big cats with slower maturing cubs and cervids with bigger antlers (Adams et al. 2016, 14). It is possible that artificial selection eventually could create a new species in the sense of creating domesticates no longer able to produce viable offspring with their wild progenitors but that process is generally a very long one (Knox 2011, 47).

Another important issue is public perception of high-fence operations. Some are literal kill pens, which repulse many people, hunters and non-hunters alike (Adams et al. 2016, 19), though they date at least to the nineteenth century (Clemens 1886). Many modern outfits, though, offer more genuine hunting experiences, replete with failure. High-fence operations vary by the degree of freedom of movement and choice that they afford game animals. There is a huge difference between merely enclosing 100,000 acres for management purposes and a 10-acre operation with artificially inseminated deer, no natural browse, and one feeder or one artificial funnel (Peterson et al. 2016, 431–32).

High-fence operations can be regulated so they do not help spread any disease or give the impression that they are more about shooting and killing than hunting (Knox 2011, 45). The game within them do not need to be managed by the government. Moreover, they need not supplant public lands, wildlife, or management. There is no reason why public and private hunting grounds cannot co-exist, especially if outdoor equipment excise taxes and tag fees remain earmarked for government conservation programs. The state can manage wildlife on state lands and on unfenced private lands under the public trust doctrine and private owners can manage wildlife on private lands under general property and business principles.

Keeping wildlife management decisions at the state level will ensure some competition between competing ideas and systems, like the starkly different approaches of Virginia and Texas. But even in the latter state, canned trophy hunts remain a small portion of the total and likely will remain so as they offer most hunters little sense of accomplishment (Knox 2011, 45). "Fair chase" rules certainly have a "cultural" (Braverman 2015, 148) component and hence are malleable. From a scientific wildlife management perspective, however, they are mostly about tradeoffs between take rates, season lengths, and cull criteria (Organ et al. 2012, 16–17; Braverman 2015, 148–49, 163).

References

Adams, Kip P., Brian P. Murphy, and Matthew D. Ross. 2016. Captive White-Tailed Deer Industry: Current Status and Growing Threat. *Wildlife Society Bulletin* 40 (1): 14–19.

Anon. 1876. Hunting and Fishing. *St. Louis Globe-Democrat* (209 Oct.), 6.

———. 1894. Hunting Wild Geese. *Milwaukee Daily Sentinel* (28 Oct.), 13.

———. 1895a. Hunters and Hunting. *Milwaukee Daily Sentinel* (3 Nov.), 11.

———. 1895b. The Hunting Paradises. *Daily Picayune* (29 Sept.), 21.

———. 1895c. Venison as Food. *Milwaukee Daily Sentinel* (29 Mar.), 4.

Baltimore Sun. 1896. Trapping Park Deer. *Emporia Daily Gazette* (16 Mar.).

Braverman, Irus. 2015. Conservation and Hunting: Till Death Do They Part? A Legal Ethnography of Deer Management. *Journal of Land Use & Environmental Law* 30 (2): 143–199.

Brennan, Leonard A., and William P. Kuvlesky. 2005. North American Grassland Birds: An Unfolding Conservation Crisis? *Journal of Wildlife Management* 69 (1): 1–13.

Brooks, Jeffrey James, and Kevin Andrew Bartley. 2016. What Is a Meaningful Role? Accounting for Culture in Fish and Wildlife Management in Rural Alaska. *Human Ecology* 44 (5): 517–531.

Budiansky, Stephen. 1992. *The Covenant of the Wild: Why Animals Chose Domestication.* New York: William Morrow.

Burroughs, R.D. 1937. An Analysis of Hunting Records for the Prairie Farm Project, Saginaw County, Michigan, 1937. *Journal of Wildlife Management* 3 (1): 19–25.

Child, Brian. 2019. *Sustainable Governance of Wildlife and Community-Based Natural Resource Management: From Economic Principles to Practical Governance.* New York: Routledge.

Clemens, Will M. 1886. Hunting Pennsylvania Deer. *Detroit Free Press* (31 July), 6.

Cummins, James L. 2018. Role of the Nonprofit Sector in Policymaking. In *North American Wildlife Policy and Law*, ed. Bruce D. Leopold, Winifred B. Kessler, and James L. Cummins. Boone and Crockett Club: Missoula.

Czech, Brian. 2000. Economic Growth as a Limiting Factor for Wildlife Conservation. *Wildlife Society Bulletin* 28 (1): 4–15.

Decker, Daniel J., John F. Organ, Ann B. Fortschen, Cynthia A. Jacobson, William F. Siemer, Christian A. Smith, Patrick E. Lederle, and Michael V. Schiavone. 2017. Wildlife Governance in the 21st Century: Will Sustainable Use Endure? *Wildlife Society Bulletin* 41 (4): 821–826.

Dickinson, Nate. 1993. *Common Sense Wildlife Management: Discourses on Personal Experiences*. Altamont, NY: Settle Hill Publishing.

Dunbar, Gary S. 1962. Deer-Keeping in Early South Carolina. *Agricultural History* 36 (2): 108–109.

Earle, Peter C. 2020. *Coronavirus and Human Rights*. Great Barrington: American Institute for Economic Research.

Erie Dispatch. 1884. Hunting in Pennsylvania. *Cleveland Daily Herald* (13 Jan.), 12.

Feldpausch-Parker, Andrea, Israel D. Parker, and Elizabeth S. Vidon. 2017. Privileging Consumptive Use: A Critique of Ideology, Power, and Discourse in the North American Model of Wildlife Conservation. *Conservation and Society* 15 (1): 33–40.

Golden, Katherine E., M. Nils Peterson, Christopher S. DePerno, Robert E. Bardon, and Christopher E. Moorman. 2013. Factors Shaping Private Landowner Engagement in Wildlife Management. *Wildlife Society Bulletin* 37 (1): 94–100.

Harris, Larry D. 1984. *The Fragmented Forest: Island Biogeography Theory and the Preservation of Biotic Diversity*. Chicago: University of Chicago Press.

Helena Independent. 1896. Hunting Elk in Montana. *Galveston Daily News* (5 June).

Holcombe, Randall. 2019. *Liberty in Peril: Democracy and Power in American History*. San Francisco: Independent Institute.

Huffman, James L. 1995. In the Interests of Wildlife: Overcoming the Tradition of Public Rights. In *Wildlife in the Marketplace: The Political Economy Forum*, ed. Terry L. Anderson and Peter J. Hill. New York: Rowman & Littlefield.

Jahn, Laurence R. 1991. Foreword. In *Wildlife and Habitats in Managed Landscapes*, ed. Jon E. Rodiek and Eric G. Bolen. Washington, DC: Island Press.

Jonker, Sandra A., Robert M. Muth, John F. Organ, Rodney R. Zwick, and William F. Siemer. 2006. Experiences with Beaver Damage and Attitudes of Massachusetts Residents Toward Beaver. *Wildlife Society Bulletin* 34 (4): 1009–1021.

Knox, W. Matt. 2011. The Antler Religion. *Wildlife Society Bulletin* 35 (1): 45–48.

Lueck, Dean L. 1995. The Economic Organization of Wildlife Institutions. In *Wildlife in the Marketplace: The Political Economy Forum*, ed. Terry L. Anderson and Peter J. Hill. New York: Rowman & Littlefield.

New York Sun. 1888. Trapping Wild Beasts. *Milwaukee Journal* (11 Sept.).

O. 1889. Deer Hunting. *Portland Oregonian* (13 Jan.).

Organ, John F. 2018. The North American Model of Wildlife Conservation. In *North American Wildlife Policy and Law*, ed. Bruce D. Leopold, Winifred B. Kessler, and James L. Cummins. Boone and Crockett Club: Missoula.

Organ, John F., et al. 2012. The North American Model of Wildlife Conservation. *The Wildlife Society Technical Review* 12-04.

Organ, John F., Thomas A. Decker, and Tanya M. Lama. 2016. The North American Model and Captive Cervid Facilities—What Is the Threat? *Wildlife Society Bulletin* 40 (1): 10–13.

Ostrom, Elinor. 1990. *Governing the Commons: The Evolution of Institutions for Collective Action*. New York: Cambridge University Press.

Palmer, E. Laurence. 1939. Farm Forest Facts. *Cornell Rural School Leaflet* 33 (2): 1–32.

Peterson, Markus J., M. Nils Peterson, and Tarla Rai Peterson. 2016. What Makes Wildlife Wild?: How Identity May Shape the Public Trust Versus Wildlife Privatization Debate. *Wildlife Society Bulletin* 40 (3): 428–435.

Schorr, Robert A., Paul M. Lukacs, and Justin A. Gude. 2014. The Montana Deer and Elk Hunting Population: The Importance of Cohort Group, License Price, and Population Demographics on Hunter Retention, Recruitment, and Population Change. *Journal of Wildlife Management* 78 (5): 944–952.

Vantassel, Stephen M., Tim L. Hiller, Kelly D.J. Powell, and Scott E. Hyngstrom. 2010. Using Advancements in Cable-Trapping to Overcome Barriers to Furbearer Management in the United States. *Journal of Wildlife Management* 74 (5): 934–939.

Vrtiska, Mark P., James H. Gammonley, Luke W. Naylor, and Andrew H. Raedeke. 2013. Economic and Conservation Ramifications from the Decline of Waterfowl Hunters. *Wildlife Society Bulletin* 37 (2): 380–388.

Wagner, Richard E. 1998. The Constitutional Protection of Private Property. In *Who Owns the Environment?* ed. Peter J. Hill and Roger E. Meiners. New York: Rowman & Littlefield.

Williams, M.C. 1889. Happy Hunting Grounds. *Daily Inter Ocean* (1 Sept.), 21.

Winkler, Richelle, and Keith Warnke. 2013. The Future of Hunting: An Age-Period-Cohort Analysis of Deer Hunter Decline. *Population and Environment* 34 (4): 460–480.

Wright, Charles. 1868. Deer and Deer-Hunting in Texas. *The American Naturalist* 2 (9): 466–476.

Proxy Hunting and Other Second-Best World Policy Proposals

Abstract Chapter 5: Suggests several ways that wildlife managers, short of reinstating markets for wild game meat, can increase the number of consumptive users and political support for the sustainable, consumptive use of wildlife. The most important suggestion is for the legalization of proxy hunting, where needed, to meet local scientific wildlife management goals, as a way to increase wild game harvests short of full-blown re-commercialization.

Keywords North American Wildlife Conservation Mode • Scientific wildlife management • Proxy hunting • Wild game hunting • Re-commercialization of markets for wild game meat

Longtime New York state wildlife manager Nate Dickinson warned that "common sense must prevail. … An open mind is essential" if wildlife are to be protected from human predation and human-induced habitat change (1993, 10, 11, 17). To the extent that majoritarian democracy remains a threat to consumptive use, policies that increase public support of fishing, hunting, and trapping without endangering wildlife sustainability bear special consideration. Consumption of game meat correlates strongly with public opinion of hunting in Sweden (Ljung et al. 2012). The relationship

© The Author(s), under exclusive license to Springer Nature
Switzerland AG 2022
R. E. Wright, *The History and Evolution of the North American
Wildlife Conservation Model*,
https://doi.org/10.1007/978-3-031-06163-9_5

could be causal (673–74), suggesting that widening the distribution of wild game meat could improve political support for hunting among non-hunters.

If, despite all the evidence that wild meat markets do not entail certain extinction (Huffman 1995, 33; Vercauteren et al. 2011; Wright 2022), the commercialization of wild game meat remains too repugnant for wildlife managers to consider, an interim step called "proxy hunting" might help to balance the number of animals harvested with scientific management goals.[1] As noted previously, hunting services called "guiding" and "outfitting" have long been common and many enterprising people stand ready to extend such services whenever and wherever possible (Anderson and Hill 1995). Guides and outfitters aid licensed hunters in the taking of game, typically "trophy" animals on land that they own or lease and often also manage. Only in exceptional circumstances, however, can guides directly harvest animals on behalf of their clients.

Proxy hunting would extend guiding services one step further by allowing guides, outfitters, and other enterprising individuals to act as proxies and fill tags without the physical presence of the tagholder. The retail sale of wild mammal meat would remain illegal but more people would pay for hunting licenses and more game would be taken by lawfully allowing hunters to hunt, tag, and process entire game animals for people who cannot, or do not wish to, do so themselves. It is not the same as selling the tag itself, which has long been verboten, because the tag holder would retain legal ownership of the harvested animal (Lueck 1995, 19).

Due to an aging population and increasing, if rather misguided (Desrochers and Shimizu 2012, 9, 12–13), interest in locavorism, more people want wild meat than are willing or able to procure it for themselves (Braverman 2015, 190). How many people will demand proxy hunting services will depend on its market price, which will depend on the interaction between that demand and the supply of the service. Many hunters will not take more deer than they know they can use themselves or give away without undue hardship (Vercauteren et al. 2011, 188). This is proof of the strong conservation ethic that the NAWCM helped to build. But some hunters would take more deer if they could be compensated for their time and effort (Vercauteren et al. 2011, 188–90). While impossible to predict prices, studies suggest that people would pay up to $1500 for a legally

---

[1] Proxy fishing and trapping do not seem necessary because of the relatively low cost of entry in the former and the low level of regulation in the latter.

taken and fully processed adult white-tailed deer delivered to their freezer (Schwabe and Schuhmann 2002, 609). Many avid hunters, especially those in low-income rural communities, would be happy to supply deer at that price.

Proxy hunting is as old and wide as humanity itself and is widespread today, even where technically illegal. It remains widely practiced, usually without monetary compensation (https://www.blm.gov/sites/blm.gov/files/uploads/Programs_Natural-Resources_Subsistence_Alaska_FactSheet.pdf), by subsistence hunters in Alaska (Brooks and Bartley 2016), the Canadian Arctic (Tobias and Kay 1994), the Faroe Islands (Shoemaker 2005, 273), and elsewhere under terms like "designated hunter."

My own mother filled deer tags almost annually for decades although she never touched a bow or gun or set foot in the woods. What my now-deceased father did was technically illegal but not unethical and I daresay that more than one game warden knew what he was doing but looked the other way. We were poor, my mother had to work, and my father was disabled but had friends, and later sons, who field dressed and dragged his harvest to the old minivan that served as his truck. He applied and paid for my mother's tag and harvested the deer in season using lawful methods; often the same day he filled his own tag. This sort of informal proxy hunting remains common, especially on farms and ranches where the risk of being caught is nearly nil.[2]

Making proxy hunting lawful would merely render the practice more transparent and allow its expansion to strangers. The number of people who would like to consume wild meat but who do not want to hunt or cannot do so for cultural, familial, or physical reasons is unknown and cannot be known with certainty until proxy hunting becomes legal. As matters stand now, they must beg cuts from hunter friends or food pantries. While hunters under the NAWCM cannot sell wild game meat, they can gift or donate it as they see fit and many do, quite generously. But hunters can give away only so much of their time and money (for tags, shells, fuel, etc.), and no state or province, to my knowledge, sells them an additional tag because they donated meat from an earlier harvest.

---

[2] Depredation tags approximate proxy hunting, but they are somewhat different in that a hunter kills nuisance animals on behalf of a landowner but keeps the meat as his reward for helping the landowner (Braverman 2015, 172, 187).

In other words, donations come out of a fixed pie of wild game meat and hence merely redistribute meat rather than satiating total demand, especially at the price of zero. Lawful proxy hunting, by contrast, would increase the number of tags purchased and the amount of wild game taken, which is precisely what is needed at present in many locales. If prices were unregulated and barriers to entry were low, market prices for proxy hunting services would soon equilibrate supply and demand. If no market forms once legal barriers are removed, no harm has been done to wildlife.

Proxy hunters would have incentives to push for looser hunting restrictions, just as commercial fishers have incentives to push for looser regulations (Biber and Eagle 2015, 788), but if limited to sole proprietors or LLCs, they would not have much sway. As long as scientific wildlife management does not become politicized, or controlled by a dominant interest group, proxy hunters will remain just one source of influence among several.

If the number of hunters cannot be increased enough to meet scientific wildlife management goals through proxy hunting, perhaps they can be by increasing the net benefits associated with hunting. Like the ancient Greeks, wildlife managers could try to make hunting about education, exercise, leisure, preparation for war or catastrophe, therapy, or status (Sokos et al. 2014). More concretely, wildlife managers could reduce the costs or increase the benefits of hunting with any number of minor policy tweaks, including, most simply, increasing the ease of hunter access to spots where they are most likely to encounter game (Cable 1991, 48–51; Lebel et al. 2012).

Hunter "age out" (Winkler and Warnke 2013, 468) can be slowed by loosening various rules. Regulations regarding placing blinds on public land, shooting from vehicles, and the like drive hunters toward "shooting preserves" where they can be dropped off right next to blinds, get help with animal removal, field dressing and processing, and so forth. Allowing gun bearers and other helpers on public land would help and could constitute an intermediate step toward proxy hunting. To break hunters of the notion that killing female deer is bad, as many still believe, education is necessary, as are incentives like "earn a second buck tag" (Braverman 2015, 181). Crossbows are safer than guns and easier to master than vertical bows, so they can aid in hunter recruitment (Braverman 2015, 182).

When wild game was getting scarce in some places at the end of the nineteenth century, states deliberately coordinated their seasons so that hunters had to pick a state and hunt it, rather than hunting sequentially in one state after another. They also began to charge out-of-state hunters higher license fees to dissuade too many from engaging in hunting arbitrage, that is, hunting in states with the lowest fees. Higher license fees also tended to dissuade less desirable hunters, those more likely to poach and less able to spend large sums on hotels and such (Johnston et al. 1896). Such measures worked, suggesting *their reversal* today could induce die-hard hunters to engage in more out-of-state hunting, especially in those areas where their help is most needed to keep wild game populations in check.

Wildlife managers and in-state sportsmen will bristle at the thought of reducing revenues by decreasing license fees for out-of-staters. But quantities purchased could be increased without lowering price simply by making the licenses more valuable to out-of-state consumptive users. It is unclear, for example, why out-of-state fishing and hunting licenses have to be based on consecutive days, or blocks of consecutive days. Simply create a phone app where consumptive users indicate that they are fishing or hunting that day and subtract it from the total number of days paid for, with fines for those who do not comply.

Similarly, consumptive users care about the risk of getting caught and fined for inadvertent "gotcha" offenses with no clear scientific purpose. The number of duck hunters, for example, is no longer driven by the number of ducks. In other words, waterfowl hunting is in secular decline despite what wildlife managers perceive as liberal rules (Vrtiska et al. 2013, 380–81). That is a problem because wetlands are going unprotected due to the decline of waterfowlers' dollars. Getting rid of the 3-shot limit for waterfowl hunters could help by eliminating the need to plug shotguns and by raising hunter harvests. Five or six shells hardly seems unsporting and those seeking greater challenge can always plug if they see fit. Similarly, limits on specific duck species makes it more difficult for hunters who have to make quick judgments, often in low-light conditions. While matters have improved since the notorious "points" system days, many would-be waterfowlers believe the current rules generate citation revenue rather than hunter utility. It is especially galling to be checked by game wardens who cannot identify duck species even by their common names when the birds are in the back of the truck! How is somebody who hunts waterfowl once or twice a season expected to identify the species of flying ducks? One idea under discussion in South Dakota and elsewhere is an "any 3 ducks" bag limit option. Another way of diminishing such "regulatory risk" is to move to season limits instead of daily bags.

In many states, more tags can only be acquired by participating in more seasons (Braverman 2015, 173), which requires buying and familiarizing oneself with new weapons and other kit and perhaps developing new hunting strategies and lands. Wildlife managers can tweak hunter success rates by adjusting season length for different weapons rather than limiting tags by weapon type. Many states already allow archers to use archery equipment to fill firearm tags, but they could do the opposite as well, allowing each hunter to decide how to fill the X tags allotted to him or her that year. Giving archers and primitive firearms hunters first crack at big bucks and full freezers, along with the gratification that use of such weapons bring, will draw many into the field, with the option to use a rifle later if they are unsuccessful.

Restrictions on weapon use are obviously necessary in some geographical areas. The same hunter who safely can have a blast with his .300 Win Mag in the woods of northern Minnesota would be a positive menace in suburban Minneapolis and other shotgun-only zones. Wildlife managers might consider, though, designating more public land archery-only for reasons of sport rather than safety. Archers tend to be quieter, more patient hunters than those who hunt with firearms and suffer costs when firearm seasons overlap with archery season, as they do in many states. Not every archer has lost a Booner to a bumbling gun-toting neophyte but many imagine they have, making them that much less likely to continue the sport, especially if firearm seasons are extended, which is perhaps the bluntest way to increase harvest totals.

Minor rule changes could also increase the benefits associated with hunting. Hunter success rates and overall satisfaction, for example, might be increased by careful consideration of regulatory mix. In South Dakota, for example, archers can hang ladder stands on public lands a full month before the deer archery season begins and can leave them up until a full month after the season ends. Only archers with a doctor-verified handicap, however, can lawfully fill archery tags with a crossbow. In Georgia, by contrast, crossbows are legal for all but placing stands on public land for more than a few days at a time is verboten. Archer success rate would undoubtedly increase in both states if both crossbow use *and* ladder stands were unrestricted as old men and many women, the fastest growing group of hunters (Schorr et al. 2014, 949), are not keen on packing in climbing stands along with the rest of their kit on a daily basis. For many, ladder stands are much safer and more comfortable, especially if put up only 12 feet.

In terms of small game, a switch to seasonal instead of daily bag limits may increase hunter numbers. Currently, entirely for historical reasons based on the ease of enforcing limits, most states limit the number of rabbits, squirrels, pheasants, quail, and other small game that a hunter may take on any given day. This increases incentives for hunters to hunt in small groups so that they may essentially share bag limits. Lone hunters resent that advantage but not as much as they resent competing against hunters who, due to residence and occupation, can hunt on more days and hence lawfully take more game without paying any more for his or her license.

Imagine a daily bag of 5 over a 100-day season. Resident hunters pay $100 for a license regardless of how often they hunt or how many animals they bag. If Hunter A hunts every day and limits out every day ($5*100 = 500$ harvests) she pays only 20 cents per animal ($100/500) in license fees. If Hunter B hunts 5 days and limits out (25 harvests), he pays $4 per animal. If Hunter C hunts 10 days but averages only 2.5 harvests per hunt, he still pays $4 per animal and, learning of Hunter A's success, feels aggrieved that she limits his success simply because she can hunt so often. He feels especially aggrieved on days when he could have harvested 10 or 20 animals were it not for the arbitrary daily limits.

Seasonal bag limits would alleviate such perceived injustices by capping every hunter at, say, 100 harvests regardless of the number of days hunted. While total harvests will remain a function of hunter skill and time allotted to hunting, Hunter C will no longer feel like locals are monopolizing game and will realize that the value of his license is higher. He could get lucky and hit the right field at the right time and fill half his season limit in a day. He and his spouse are much happier than if he comes home with only 5 and stories about how he could have harvested many more.

Daily bag limits arose because it was the cheapest and easiest way to monitor harvests under the old adage, "if the duck is in the truck." Some states, like Michigan, tried both daily and seasonal bag limits for highly sought game like pheasants, enforcing both with chokepoint game check stations (Burroughs 1937, 21 n.2). Some states switched to such low season limits on small game, like ten pheasants per season in Connecticut (which also imposes a 2 per day limit), that they could adopt a big game-like tag system. Physically tagging the million or so pheasants taken annually in South Dakota, or the half a million squirrels (*Sciurus carolinensis* and *Sciurus niger*) harvested in Illinois, would be prohibitively costly. Modern technology, though, is currently being used in Georgia to "tag" deer, and there is no reason why it could not be extended even to doves or snow geese. Instead of showing their daily bag to the game warden, hunters have to show that they have recorded their kills electronically using a smart phone. Modern apps, like that of Georgia, work even without cell phone service, storing the data until a telephone or internet connection is made. In addition to keeping a tally of the hunter's seasonal bag, the system could be used to collect all sorts of information useful to wildlife managers, including the precise location and time of the kill, the weapon used, and so forth. The app could serve as a sort of hunter's diary and could automatically upload pictures to social media. It could also be used to self-report game violations without the embarrassment of face-to-face interaction with a game warden.

Finally, wildlife managers need to keep hunting a human activity. While acceptable equipment can be managed to increase or decrease harvest rates at the scientifically informed discretion of wildlife managers, allowing machines to do most of the work, especially pulling the trigger, could quickly become problematic. Hunting by proxy, in other words, ought not mean hunting by robot. That may become an important point in the future, as AI-assisted robotics becomes more sophisticated and capable. If hunting and trapping remain the sole purview of flawed, lazy humans, most animal species will be able to survive their predation, holding habitat constant of course. Moreover, tying proxy hunting to jobs can help voters to see it as legitimate (Decker et al. 2017, 85), as tying jobs to private wildlife management did after the shift to privatized sustainable use regimes in Africa following the 1963 Arusha Conference (Child et al. 2012).

REFERENCES

Anderson, Terry L., and Peter J. Hill. 1995. From a Liability to an Asset: Developing Markets for Wildlife. In *Wildlife in the Marketplace: The Political Economy Forum*, ed. Terry L. Anderson and Peter J. Hill. New York: Rowman & Littlefield.

Biber, Eric, and Josh Eagle. 2015. When Does Legal Flexibility Work in Environmental Law? *Ecology Law Quarterly* 42 (4): 787–840.

Braverman, Irus. 2015. Conservation and Hunting: Till Death Do They Part? A Legal Ethnography of Deer Management. *Journal of Land Use & Environmental Law* 30 (2): 143–199.

Brooks, Jeffrey James, and Kevin Andrew Bartley. 2016. What Is a Meaningful Role? Accounting for Culture in Fish and Wildlife Management in Rural Alaska. *Human Ecology* 44 (5): 517–531.

Burroughs, R.D. 1937. An Analysis of Hunting Records for the Prairie Farm Project, Saginaw County, Michigan, 1937. *Journal of Wildlife Management* 3 (1): 19–25.

Cable, Ted T. 1991. Windbreaks, Wildlife, and Hunters. In *Wildlife and Habitats in Managed Landscapes*, ed. Jon E. Rodiek and Eric G. Bolen. Washington, DC: Island Press.

Child, Brian, Jessica Musengezi, Gregory D. Parent, and Graham F.T. Child. 2012. The Economics and Institutional Economics of Wildlife on Private Land in Africa. *Pastoralism: Research, Policy and Practice* 18 (2): 1–32.

Decker, Daniel J., John F. Organ, Ann B. Fortschen, Cynthia A. Jacobson, William F. Siemer, Christian A. Smith, Patrick E. Lederle, and Michael V. Schiavone. 2017. Wildlife Governance in the 21st Century: Will Sustainable Use Endure? *Wildlife Society Bulletin* 41 (4): 821–826.

Desrochers, Pierre, and Hiroko Shimizu. 2012. *The Locavore's Dilemma: In Praise of the 10,000-Mile Diet*. New York: Public Affairs.

Dickinson, Nate. 1993. *Common Sense Wildlife Management: Discourses on Personal Experiences*. Altamont, NY: Settle Hill Publishing.

Huffman, James L. 1995. In the Interests of Wildlife: Overcoming the Tradition of Public Rights. In *Wildlife in the Marketplace: The Political Economy Forum*, ed. Terry L. Anderson and Peter J. Hill. New York: Rowman & Littlefield.

Johnston, John, J.W. Ormsby, H.N. Campbell, Edward Silverman, H.T. Drake, Stephen Meunier, C.H. Mathews, Alfred James, and R.G. Richter. 1896. The Hunting of Deer. *Milwaukee Journal* (7 Nov.).

Lebel, Francois, Christian Dussault, Ariane Masse, and Steve D. Cote. 2012. Influence of Habitat Features and Hunter Behavior on White-Tailed Deer Harvest. *Journal of Wildlife Management* 76 (7): 1431–1440.

Ljung, Per E., Shawn J. Riley, Thomas A. Heberlein, and Go Ran Ericsson. 2012. Eat Prey and Love: Game-Meat Consumption and Attitudes Toward Hunting. *Wildlife Society Bulletin* 36 (4): 669–675.

Lueck, Dean L. 1995. The Economic Organization of Wildlife Institutions. In *Wildlife in the Marketplace: The Political Economy Forum*, ed. Terry L. Anderson and Peter J. Hill. New York: Rowman & Littlefield.

Schorr, Robert A., Paul M. Lukacs, and Justin A. Gude. 2014. The Montana Deer and Elk Hunting Population: The Importance of Cohort Group, License Price, and Population Demographics on Hunter Retention, Recruitment, and Population Change. *Journal of Wildlife Management* 78 (5): 944–952.

Schwabe, Kurt A., and Peter W. Schuhmann. 2002. Deer-Vehicle Collisions and Deer Value: An Analysis of Competing Literatures. *Wildlife Society Bulletin* 30 (2): 609–615.

Shoemaker, Nancy. 2005. Whale Meat in American History. *Environmental History* 10 (2): 269–294.

Sokos, Christos K., M. Nils Peterson, Periklis K. Birtsas, and Nikolas D. Hasanagas. 2014. Insights for Contemporary Hunting from Ancient Hellenic Culture. *Wildlife Society Bulletin* 38 (3): 451–457.

Tobias, Terry N., and James J. Kay. 1994. The Bush Harvest in Pinehouse, Saskatchewan, Canada. *Arctic* 47 (3): 207–221.

Vercauteren, Kurt C., Charles W. Anderson, Timothy R. Van Deelen, W. David Drake, David Walter, Stephen M. Vantassel, and Scott E. Hyngstrom. 2011. Regulated Commercial Harvest to Manage Overabundant White-Tailed Deer: An Idea to Consider? *Wildlife Society Bulletin* 35 (3): 185–194.

Vrtiska, Mark P., James H. Gammonley, Luke W. Naylor, and Andrew H. Raedeke. 2013. Economic and Conservation Ramifications from the Decline of Waterfowl Hunters. *Wildlife Society Bulletin* 37 (2): 380–388.

Winkler, Richelle, and Keith Warnke. 2013. The Future of Hunting: An Age-Period-Cohort Analysis of Deer Hunter Decline. *Population and Environment* 34 (4): 460–480.

Wright, Robert E. 2022. The Political Economy of Modern Wildlife Management: How Commercialization Could Reduce Game Overabundance. *Independent Review* 26 (4): 512–32.

CHAPTER 6

Conclusion

Abstract This chapter concludes that scientific wildlife management, checked by the need to maintain inflows of consumptive user dollars, will protect North America's wildlife throughout the twenty-first century better than meat market bans can.

Keywords North American Wildlife Conservation Model • Scientific wildlife management • Consumptive user • Political economy of conservation • Re-commercialization of markets for wild game meat • Proxy hunting

Early in the twentieth century, the piecemeal adoption of a set of policies later known as the North American Wildlife Conservation Model (NAWCM) helped to save many wild game species in North America from extensive range restriction and possibly even permanent obliteration, though economic forces would probably have interceded as they did for bison (privatized), caribou (managed), beaver and moose (too expensive to trap or hunt as their numbers dwindled), and other animals. All seven of the model's main components, the Seven Sisters, contributed to a dramatic reversal in the fortunes of game animals like white-tailed deer and

© The Author(s), under exclusive license to Springer Nature
Switzerland AG 2022
R. E. Wright, *The History and Evolution of the North American
Wildlife Conservation Model*,
https://doi.org/10.1007/978-3-031-06163-9_6

wild turkeys by breaking customs and hunting practices detrimental to their long-term sustainability in an increasingly high technology, high population modern world.

More licensing, fencing, trespass law enforcement, and federal and international treaties reduced the tragedy of the commons/common pool problem. Poaching decreased with increased wildlife law enforcement and cheaper, more uniform farmed alternative sources of animal protein, fats, and furs. Furthermore, less poaching led to even less poaching as hunters realized that most hunters were following the new rules so there was no race to kill everything before other hunters did.

Attitudes changed too, as hunting became more about sport and the quality of the animals harvested than about quantity. Restrictions on when, where, and how animals could be harvested, like banning night hunting for many species, greatly reduced harvest rates per hunter while simultaneously increasing the enjoyment of the hunt for many outdoorsmen increasingly attracted to the evolving rules of "fair chase." Moreover, banning the sale of wild game meat reduced incentives to harvest, allowing the new restrictions to normalize and inducing commercial hunters to migrate into guiding and outfitting rather than killing everything that moved, as was long customary in North America.

A century after the development of the NAWCM, many places in North America now face a new challenge, the local overabundance of deer, wild hogs, horses, turkeys, and other species. The number of hunters has dropped in many areas, and they simply do not harvest enough animals to keep them in check. Reintroducing wolves into highly populated areas will not work and other cull methods remain ineffective or prohibitively expensive.

One ongoing response to game animal overabundance are attempts to increase the number of hunters and some strides in the recruitment of female hunters have been made. Loosening some regulations could help as well. Another approach is to induce avid hunters to harvest more animals by allowing them to serve as paid proxy hunters for those who cannot hunt for themselves.

Ultimately, however, North American wildlife managers may have to allow commercial harvests once again, repugnant as that may seem to some steeped in the concretized mythology of the NAWCM (Wright 2022). As explained in detail above, markets for wild meat did not cause extinctions or near extinctions; combinations of factors, including the tragedy of the commons, pathogens, new technologies, and the herding/

flocking behaviors of some species, did. Successful commercial fisheries and alligator hunting suggest that commercial hunting combined with scientifically determined harvest limits can achieve management goals without endangering game species.

The experience of other nations, like South Africa and Sweden, also suggests that commercial harvest of game animals can be undertaken in a responsible manner and can actually improve wildlife quality by meeting scientific management goals more closely than sport hunting alone may. As the history of some fisheries and whaling management show, the main threat to wildlife is not from the markets themselves but from the capture of wildlife managers by commercial interests and the subsequent increase of harvest limits over sustainable levels. Once made aware of that threat, sport hunters, nonprofits, and environmental groups will surely monitor the situation, and as long as the underlying institutional framework contains sufficient checks against such rent-seeking activities, no repeat of the late nineteenth century will occur.

REFERENCE

Wright, Robert E. 2022. The Political Economy of Modern Wildlife Management: How Commercialization Could Reduce Game Overabundance. *Independent Review* 26 (4): 512–32.

References

A Hunter. 1881. A Hunting Party. *Arkansas Gazette* (31 Dec.)
Adams, Kip P., Brian P. Murphy, and Matthew D. Ross. 2016. Captive White-Tailed Deer Industry: Current Status and Growing Threat. *Wildlife Society Bulletin* 40 (1): 14–19.
Altoona Tribune. 1886. A Hunter Rescued by His Dog. *St. Louis Globe-Democrat* (21 Dec.), 8.
Anderson, Terry L. 1995. *Sovereign Nations or Reservations? Indian Economies: An Economic History of American Indians*. Pacific Research Institute.
———. 1998. Viewing Wildlife Through Coase-Colored Glasses. In *Who Owns the Environment?* ed. Peter J. Hill and Roger E. Meiners, 259–282. New York: Rowman & Littlefield.
Anderson, Terry L., and Peter J. Hill. 1989. *The Birth of a Transfer Society*. New York: University Press of America.
———. 1995. From a Liability to an Asset: Developing Markets for Wildlife. In *Wildlife in the Marketplace: The Political Economy Forum*, ed. Terry L. Anderson and Peter J. Hill. New York: Rowman & Littlefield.
Anderson, Terry L., and Donald R. Leal. 2001. *Free Market Environmentalism*, Rev. ed. New York: Palgrave.
Anderson, Terry L., and Gary D. Libecap. 2014. *Environmental Markets: A Property Rights Approach*. New York: Cambridge University Press.
Angrist, Joshua, Pierre Azoulay, Glenn Ellison, Ryan Hill, and Susan Feng Li. 2017. Economic Research Evolves: Fields and Styles. *American Economic Review: Papers & Proceedings* 107 (5): 293–297.
Anon. 1819. Great Hunting. *National Intelligencer* (25 Nov.).

© The Author(s), under exclusive license to Springer Nature Switzerland AG 2022
R. E. Wright, *The History and Evolution of the North American Wildlife Conservation Model*,
https://doi.org/10.1007/978-3-031-06163-9

———. 1842. Walton and Cotton's Complete Angler.... *North American Review* 55 (117): 343–372.

———. 1844. Railroad Venison. *National Intelligencer* (27 Nov.).

——— 1849. City Brickbats and Pebbles, Picked Up in the Streets. *Daily Scioto Gazette* (22 Dec.).

———. 1854a. Bear Hunting. *The Illustrated Magazine of Art* 4 (20): 99–100.

———. 1854b. Wild Pigeons. *Daily South Carolinian* (13 Oct.).

———. 1858. Steam Squirrel Hunting. *Scientific American* 13 (30): 233.

———. 1860. The Sioux Indians. *Omaha Nebraskan* (11 Feb.).

———. 1862. Frog Hunting. *Scientific American* 7 (3): 35.

———. 1866a. Thirty Years of Army Life ... by R. B. Marcy. *North American Review* 103 (213): 80–82.

———. 1866b. The Marten Trappers. *Frank Leslie's Illustrated Newspaper* (13 Oct.), 60.

———. 1866c. Trapping Martens in Canada. *Frank Leslie's Illustrated Newspaper* (8 Dec.), 188.

———. 1866d. Trapping the Beaver. *Daily Evening Bulletin* (29 May).

———. 1866e. We Have Heard an Estimate. *Milwaukee Daily Sentinel* (18 Jan.).

———. 1867. A Very Homely But Readable Book. *Frank Leslie's Illustrated Newspaper* (14 Dec.), 194.

———. 1869. Pigeon Trapping. *Wisconsin State Register* (24 April).

———. 1870a. Wild Life in the West. *Milwaukee Daily Sentinel* (10 Feb.).

———. 1870b. Indian Trappers. *Daily Evening Bulletin* (21 Feb.).

———. 1870c. Venison in Market. *Daily Evening Bulletin* (2 July).

———. 1871. The Winnipeg Trappers. *Independent Statesman* (23 Nov.), 59.

———. 1872. The St. Paul Press Is Informed. *Milwaukee Daily Journal* (13 Nov.).

———. 1873a. Buffalo Hunting. *Daily Graphic* (21 July), 134.

———. 1873b. The Fur Trade of Lewiston. *Lowell Daily Citizen* (23 Dec.).

———. 1874a. The Price Paid to Trappers. *Daily Evening Bulletin* (14 Jan.).

———. 1874b. The Sale of Venison to Be Stopped. *Cleveland Daily Herald* (2 Dec.), 8.

———. 1876a. A Bear Hunting Exploit. *St. Louis Globe-Democrat* (13 Aug.), 12.

———. 1876b. Hunting and Fishing. *St. Louis Globe-Democrat* (209 Oct.), 6.

———. 1876c. There Appears to Be Quite a Rivalry. *Arizona Miner* (29 Dec.).

———. 1877. A New Hunting Ground. *Scientific American* 36 (4): 49.

———. 1878. Contraband Venison. *Daily Evening Bulletin* (3 Aug.).

———. 1879. Hunting Wild Geese with Oxen. *Daily Inter Ocean* (8 Dec.), 6.

———. 1881a. Michigan Frogs and Frog Hunting. *Scientific American* 45 (13): 197.

———. 1881b. Hunting Alligators in Florida. *Scientific American* 45 (18): 297.

———. 1881c. Hunting Wild Honey. *Frank Leslie's Illustrated Newspaper* (1 Oct.), 71.

———. 1881d. Wild Pigeons for the Sportsmen's Tournament. *Frank Leslie's Illustrated Newspaper* (2 July), 299.

———. 1881e. The Mountains of Tennessee. *News and Observer* (1 Jan.).

———. 1881f. Preparing to Sacrifice Game. St. Louis Globe-Democrat (14 Oct.), 7.

———. 1882. Hunting Wild Turkeys. *Daily Inter Ocean* (9 Feb.), 5.

———. 1883a. Hunting Stories. *St. Louis Globe-Democrat* (15 Dec.), 16.

———. 1883b. The Bellefontaine Trappers. *St. Louis Globe-Democrat* (17 Feb.), 9.

———. 1883c. At Red Bluff the Sky Is Clouded. *Los Angeles Times* (6 Sept.).

———. 1884a. Trapping in Wisconsin. *St. Louis Globe-Democrat* (17 Oct.), 4.

———. 1884b. Report Says. *Galveston Daily News* (23 Nov.), 3.

———. 1885a. Fish Trapping. *Idaho Avalanche* (4 Apr.).

———. 1885b. A Muskrat's Perils. *Wisconsin State Register* (26 Dec.).

———. 1886a. Deer Hunting with Steam. *Scientific American* 55 (23): 354.

———. 1886b. Two Hunting Stories. *Detroit Free Press* (15 May), 8.

———. 1886c. Going Fox-Hunting. *Daily Inter Ocean* (11 Nov.), 3.

———. 1886d. His Hunting Trips. *Milwaukee Daily Sentinel* (2 Apr.), 4.

———. 1886e. D.P. Graves, the Well-Known Hunter. *Milwaukee Daily Sentinel* (29 Nov.), 8.

———. 1887. Costly Venison. *Daily Picayune* (10 Oct.), 4.

———. 1888. A Border Trapper's Life. *Rocky Mountain News* (11 Nov.), 23.

———. 1889. Hunting Bear in Louisiana. *Daily Inter Ocean* (26 May), 12.

———. 1890. Some Tall Hunting Stories. *Milwaukee Journal* (25 Oct.), 6.

———. 1891a. Squirrel Hunting. *Atchison Daily Champion* (16 Sept.), 2.

———. 1891b. Hunting the Wary Moose. *Daily Inter Ocean* (29 Nov.), 29.

———. 1891c. Hunting Adventures. *Bangor Daily Whig and Courier* (28 Oct.).

———. 1891d. Shipping Venison in Coffins. *Milwaukee Journal* (12 Nov.).

———. 1892. Big Game Seizure. *St. Paul Daily News* (17 Feb.).

———. 1893a. A Mighty Hunter from the South. *Daily Picayune* (9 Sept.).

———. 1893b. Trapping Beaver. *Boston Daily Advertiser* (23 Mar.), 8.

———. 1893c. Will Live on Venison. *Rocky Mountain News* (9 Sept.), 3.

———. 1894a. Hunting Wild Geese. *Milwaukee Daily Sentinel* (28 Oct.), 13.

———. 1894b. Hunting with Dogs. *Milwaukee Journal* (7 Nov.), 5.

———. 1894c. The Mightiest Hunter. *Atchison Globe* (27 Apr.), 3.

———. 1894d. Hunting Horses. *Vermont Watchman* (9 May), 2.

———. 1894e. Markets Full of Game. *Portland Oregonian* (31 Oct.), 5.

———. 1895a. Hunting Egrets in Mexico. *Daily Inter Ocean* (2 Dec.), 3.

———. 1895b. Hunters and Hunting. *Milwaukee Daily Sentinel* (3 Nov.), 11.

———. 1895c. The Hunting Paradises. *Daily Picayune* (29 Sept.), 21.

———. 1895d. Private Secretary Thurber's Zeal. *Daily Inter Ocean* (17 June), 6.

———. 1895e. Moose Hunting. *Bangor Daily Whig and Courier* (3 Dec.).

———. 1895f. Is a Brave Hunter. *Daily Inter Ocean* (28 Nov.), 7.

———. 1895g. Antelope Trapping in the West. *Daily Picayune* (2 Nov.), 12.

———. 1895h. Muskrats Are Sly Animals. *Emporia Daily Gazette* (1 March).

———. 1895i. Venison Coming In. *Rocky Mountain News* (30 Nov.), 8.

———. 1895j. Game Birds and Venison. *Bismarck Tribune* (31 Dec.).

———. 1895k. Venison as Food. *Milwaukee Daily Sentinel* (29 Mar.), 4.

———. 1896a. Duck Hunting at Home. *Milwaukee Journal* (27. Nov.).

———. 1896b. A Chicago Hunter. *Galveston Daily News* (4 Dec.), 8.

———. 1896c. A Frog Hunter. *Emporia Daily Gazette* (12 June).

———. 1896d. Trapping in Maine. *Daily Inter Ocean* (17 Apr.), 10.

———. 1896e. Raccoons in Louisiana. *Portland Oregonian* (19 Jan.), 15.

———. 1896f. Loaded with Venison. *Rocky Mountain News* (11 Jan.), 4.

———. 1896g. Tons of Venison. *Bismarck Tribune* (5 Dec.).

———. 1897a. Riparian Rights. Navigable Waters. Trespass. Rights of Hunters. Hall v. Alford, 72 N.W. Rep. 137 (Mich.). *Yale Law Journal* 7 (2): 96.

———. 1897b. Flamingo Hunting. *New Orleans Daily Picayune* (16 Aug.), 7.

———. 1897c. In Conversation. *Bangor Daily Whig and Courier* (16 Oct.)

———. 1898a. Danger in New Hunting Rifles. *Scientific American* 79 (6): 85.

———. 1898b. Hunting and Fishing. *Milwaukee Journal* (4 Nov.), 4.

———. 1898c. It Will Stop 'Pot Hunting'. *Milwaukee Journal* (14 Feb.), 6.

———. 1898d. Hunting in Maine. *Atchison Globe* (5 Oct. 1898), 3.

———. 1898e. Sold Confiscated Venison. *Milwaukee Daily Sentinel* (27 Nov.), 12.

———. 1898f. For Violating Game Law. *Milwaukee Daily Sentinel* (19 Feb.), 10.

———. 1898g. Partridges and Venison at Lynch's. *Bangor Daily Whig and Courier* (17 Nov.), 3.

———. 1899a. Coon Hunting. *Bangor Whig and Courier* (13 Oct.), 3.

———. 1899b. License for Hunting. *Milwaukee Journal* (21 July), 8.

———. 1899c. The Hunting Season. *Bangor Daily Whig and Courier* (20 Oct.), 8.

———. 1899e. A Woman Bear Hunter. *Boston Daily Advertiser* (13 June), 8.

———. 1899g. Little Venison Yet. *Milwaukee Daily Sentinel* (7 Nov.), 5.

———. 1899h. Venison for Boston Seized. *Boston Daily Advertiser* (4 Mar.), 3.

———. 1899i. Will Not Handle Venison. *Bangor Daily Whig and Courier* (16 Sept.), 8.

———. 1901. Fall Hunting in the Maine Woods. *Journal of Education* 54 (15): 246.

———. 1902. The Hunting Season Now at Its Height—Deer and Moose Very Plentiful This Year in Maine. *Journal of Education* 56 (17): 291.

———. 1909. Interference with Hunting Rights. *Harvard Law Review* 22 (4): 305–306.

———. 1913. Fox-Hunting in America. *Lotus Magazine* 5 (1): 51–67.

———. 1918. Camouflage in Seal Hunting. *Scientific American* 118 (17): 383.

———. 1923. My Life with the Eskimo…. *Geographical Teacher* 12 (3): 218–219.

———. 1925a. Handbook of Alaska…. *Washington Historical Quarterly* 16 (4): 306–308.

———. 1925b. Notorious Wolf Killed by Government Hunters. *Science News-Letter* 6 (203): 6.

———. 1932. Lead Shot Kills Ducks Even Though Hunters Fire and Miss. *Science News-Letter* 21 (566): 107.

———. 1899d. Greatest Wolf Hunter. *Atchison Globe* (15 Aug.), 3.

———. 1899f. Good Muskrat Trapping. *Atchison Globe* (10 Nov.), 3.

Archibald, Malcolm. 2013. *The Dundee Whaling Fleet: Ships, Masters and Men*. Edinburgh: Edinburgh University Press.

Audobon's Ornithological Biography. 1838. Deer Hunting. *National Intelligencer* (10 Sept.).

Baden, John A. 1998. A New Primer for the Management of Common-Pool Resources and Public Goods. In *Managing the Commons*, ed. John A. Baden and Douglas S. Noonan, 2nd ed., 51–62. Bloomington: Indiana University Press.

Baden, John A., Richard Stroup, and Walter Thurman. 1981. Myths, Admonitions, and Reality: The American Indian as a Resource Manager. *Economic Inquiry* 19 (1): 132–143.

Baltimore American. 1887. Wild Duck Trapping. *St. Louis Globe-Democrat* (5 Nov.), 15.

Baltimore Sun. 1883. Remarkable Duck Shooting. *St. Louis Globe-Democrat* (10 Nov.), 4.

———. 1892. A Cruel Method of Hunting. *Bismarck Tribune* (24 Jan.), 3.

———. 1896. Trapping Park Deer. *Emporia Daily Gazette* (16 Mar.).

Bangor Whig. 1873. Large Numbers of Deer. *Lowell Daily Citizen* (31 Jan.).

Baraboo, Wisconsin. 1899. Hunting the Wild Bee. *Milwaukee Daily Sentinel* (15 Jan.), 11.

Barboza, Perry S., and Daniel Tihanyi. 2018. State Wildlife Policy in a National Environment. In *North American Wildlife Policy and Law*, ed. Bruce D. Leopold, Winifred B. Kessler, and James L. Cummins. Boone and Crockett Club: Missoula.

Barker, John. 1896. Frozen to Death While Out Hunting. *New Orleans Daily Picayune* (30 Nov.).

Barnes, Jonathan I., James Macgregor, and L. Chris Weaver. 2002. Economic Efficiency and Incentives for Change with Namibia's Community Wildlife Use Initiatives. *World Development* 30 (4): 667–681.

Bartram, John, and Francis Harper. 1942. Diary of a Journey Through the Carolinas, Georgia, and Florida from July 1, 1765 to April 10, 1766. *Transactions of the American Philosophical Society* 33 (1): 1–120.

Basket, James Newton. 1893. Squirrel Hunting. *Atchison Globe* (1 May).

Batty, J.H. 1874. Antelope-Hunting. *The Aldine* 7 (2): 38.

Baumgartner, F.M. 1942. An Analysis of Waterfowl Hunting at Lake Carl Blackwell, Payne County, Oklahoma, for 1940. *Journal of Wildlife Management* 6 (1): 83–91.

Berthel, Mary Wheelhouse. 1935. Hunting in Minnesota in the Seventies. *Minnesota History* 16 (3): 259–271.

Biber, Eric, and Josh Eagle. 2015. When Does Legal Flexibility Work in Environmental Law? *Ecology Law Quarterly* 42 (4): 787–840.

Bish, Robert L. 1998. Environmental Resource Management: Public or Private? In *Managing the Commons*, ed. John A. Baden and Douglas S. Noonan, 2nd ed., 65–75. Bloomington: Indiana University Press.

Bissonette, John A., Richard J. Frederickson, and Brian J. Tucker. 1991. American Marten: A Case for Landscape-level Management. In *Wildlife and Habitats in Managed Landscapes*, ed. Jon E. Rodiek and Eric G. Bolen. Washington, DC: Island Press.

Blackwell, Jack. 2018. Policy and Law Relating to Tribal Wildlife Management. In *North American Wildlife Policy and Law*, ed. Bruce D. Leopold, Winifred B. Kessler, and James L. Cummins. Boone and Crockett Club: Missoula.

Bosselmann, Klaus. 2015. *Earth Governance: Trusteeship of the Global Commons*. New York: Edward Elgar.

Boston Journal. 1886. Extraordinary Duck-Hunting. *Daily Evening Bulletin* (10 Mar.), 4.

Branch, E. Douglas. 1929. *The Hunting of the Buffalo*. New York: D. Appleton and Co.

Brasseaux, Carl A., H. Dickson Hoese, and Thomas C. Michot. 2004. Pioneer Amateur Naturalist Louis Judice: Observations on the Fauna, Flora, Geography, and Agriculture of the Bayou Lafourche Region, Louisiana, 1772–1786. *Louisiana History: The Journal of the Louisiana Historical Association* 45 (1): 71–103.

Braverman, Irus. 2015. Conservation and Hunting: Till Death Do They Part? A Legal Ethnography of Deer Management. *Journal of Land Use & Environmental Law* 30 (2): 143–199.

Brennan, Leonard A., and William P. Kuvlesky. 2005. North American Grassland Birds: An Unfolding Conservation Crisis? *Journal of Wildlife Management* 69 (1): 1–13.

Brook, Barry W., and Corey J.A. Bradshaw. 2006. Strength of Evidence for Density Dependence in Abundance Time Series of 1198 Species. *Ecology* 87: 1445–1451.

Brooks, Jeffrey James, and Kevin Andrew Bartley. 2016. What Is a Meaningful Role? Accounting for Culture in Fish and Wildlife Management in Rural Alaska. *Human Ecology* 44 (5): 517–531.

Brown, Robert D. 2016. The Politics of Deer-Farming in North Carolina— Lessons Learned. *Wildlife Society Bulletin* 40 (1): 20–24.

———. 2018. The Need for Wildlife Conservation and Policy. In *North American Wildlife Policy and Law*, ed. Bruce D. Leopold, Winifred B. Kessler, and James L. Cummins. Boone and Crockett Club: Missoula.

Bryant, Fred C. 1991. Managed Habitats for Deer in the Woodlands of West Texas. In *Wildlife and Habitats in Managed Landscapes*, ed. Jon E. Rodiek and Eric G. Bolen. Washington, DC: Island Press.

Budiansky, Stephen. 1992. *The Covenant of the Wild: Why Animals Chose Domestication*. New York: William Morrow.

Burroughs, Wilbur G. 1915. Hunting in the Artic and Alaska by E. Marshall Scull. *Bulletin of the American Geographical Society* 47 (2): 139.

Burroughs, R.D. 1937. An Analysis of Hunting Records for the Prairie Farm Project, Saginaw County, Michigan, 1937. *Journal of Wildlife Management* 3 (1): 19–25.

Burroughs, R.D., and Laurence Dayton. 1941. Hunting Records for the Prairie Farm, Saginaw County, Michigan, 1937–1939. *Journal of Wildlife Management* 5 (2): 159–167.

Cable, Ted T. 1991. Windbreaks, Wildlife, and Hunters. In *Wildlife and Habitats in Managed Landscapes*, ed. Jon E. Rodiek and Eric G. Bolen. Washington, DC: Island Press.

Callaghan, Des A., Jeff S. Kirby, and Baz Hughes. 1997. The Effects on Recreational Waterfowl Hunting on Biodiversity. In *Harvesting Wild Species: Implications for Biodiversity Conservation*, ed. Curtis H. Freese. Baltimore: Johns Hopkins University Press.

Campbell, Henry C. 1889a. The Hunting Season. *Yenowine's Illustrated News* (20 Oct.), 1.

———. 1889b. Hunting Noble Game. *Yenowine's Illustrated News* (27 Oct.), 1.

Caplan, Bryan. 2007. *The Myth of the Rational Voter: Why Democracies Choose Bad Policies*. Princeton: Princeton University Press.

Carlarne, Cinnamon Pinon. 2005. Saving the Whales in the New Millennium: International Institutions, Recent Developments and the Future of International Whaling Policies. *Virginia Environmental Law Journal* 24 (1): 1–48.

Carley, Ira. 1897. Hunting Deer with Dogs. *Milwaukee Daily Sentinel* (11 Feb.), 7.

Carlos, Ann M., and Frank D. Lewis. 1995. Strategic Pricing in the Fur Trade: The Hudson's Bay Company, 1700–1763. In *Wildlife in the Marketplace: The Political Economy Forum*, ed. Terry L. Anderson and Peter J. Hill. New York: Rowman & Littlefield.

Carruthers, Jane. 2008. 'Wilding the Farm or Farming the Wild?' The Evolution of Scientific Game Ranching in South Africa from the 1960s to the Present. *Transactions of the Royal Society of South Africa* 63 (2): 160–181.

Cat, Jordi. 2017. The Unity of Science. In Edward N. Zalta, ed. *The Stanford Encyclopedia of Philosophy*. https://plato.stanford.edu/archives/fall2017/entries/scientific-unity/.

Cathlamet Gazette. 1891. Trapping Beaver. *Atchison Daily Champion* (13 Mar.), 7.

Charlotte News. 1894. A Trapping Company. *News and Observer* (13 Mar.).

Chicago Record. 1896. The Trapper's Life. *Galveston Daily News* (18 Aug.), 8.

Chicago Times. 1888. Hunting in the Rockies. *Louisville Courier-Journal* (12 Nov.), 6.

Chico Chronicle. 1886. Fur-Bearing Animals. *Los Angeles Times* (14 Feb.), 2.

Child, Brian. 2012. The Sustainable Use Approach Could Save South Africa's Rhinos. *South Africa Journal of Science* 108 (7–8): 1–4.

———. 2019. *Sustainable Governance of Wildlife and Community-Based Natural Resource Management: From Economic Principles to Practical Governance.* New York: Routledge.

Child, Brian, Jessica Musengezi, Gregory D. Parent, and Graham F.T. Child. 2012. The Economics and Institutional Economics of Wildlife on Private Land in Africa. *Pastoralism: Research, Policy and Practice* 18 (2): 1–32.

Clapham, Phillip J. 2016. Managing Leviathan: Conservation Challenges for the Great Whales in a Post-Whaling World. *Oceanography* 29 (3): 214–225.

Clayton, John. 1694. A Continuation of Mr. John Clayton's Account of Virginia. *Philosophical Transactions* 18: 121–135.

Clemens, Will M. 1886. Hunting Pennsylvania Deer. *Detroit Free Press* (31 July), 6.

Clothier and Furnisher. 1890. Trapping Muskrat. *Atchinson Daily Champion* (10 Aug.), 3.

Cooper, John M. 1929. Canadian Indians Live by Hunting. *Science News-Letter* 16 (448): 286–287.

———. 1939. Is the Algonquian Family Hunting Ground System Pre-Columbian. *American Anthropologist* 41 (1): 66–90.

Cor. New York Herald. 1889. Hunting Wild Turkeys. *Atchison Daily Champion* (6 Jan. 1889), 8.

Couzens, Ed, Alexander Paterson, and Sophie Riley. 2017. *Protecting Forest and Marine Biodiversity: The Role of Law.* New York: Edward Elgar.

Crossways, Diana. 1896. Stag Hunting. *Portland Oregonian* (10 May), 15.

Cummins, James L. 2018. Role of the Nonprofit Sector in Policymaking. In *North American Wildlife Policy and Law,* ed. Bruce D. Leopold, Winifred B. Kessler, and James L. Cummins. Boone and Crockett Club: Missoula.

Curtis, H.S. 1914. Hunting as Education. *Journal of Education* 80 (10): 260–261.

Czech, Brian. 2000. Economic Growth as a Limiting Factor for Wildlife Conservation. *Wildlife Society Bulletin* 28 (1): 4–15.

Davis, John. 1817. *Personal Adventures and Travels Four Years and a Half in the United States of America.* London: W. McDowall.

Decker, Daniel J., John F. Organ, Ann B. Fortschen, Cynthia A. Jacobson, William F. Siemer, Christian A. Smith, Patrick E. Lederle, and Michael V. Schiavone. 2017. Wildlife Governance in the 21st Century: Will Sustainable Use Endure? *Wildlife Society Bulletin* 41 (4): 821–826.

DeLong, Robert A., and Brian D. Taras. 2009. *Moose Trend Analysis User's Guide.* Alaska Department of Fish and Game.

Denver News. 1883. Riding an Elk. *St. Louis Globe-Democrat* (8 Dec.), 16.

Despain, Don, Douglas Houston, Mary Meagher, and Paul Schullery. 1986. *Wildlife in Transition: Man and Nature on Yellowstone's Northern Range.* Boulder: Roberts Rinehart, Inc.

Desrochers, Pierre, and Hiroko Shimizu. 2012. *The Locavore's Dilemma: In Praise of the 10,000-Mile Diet.* New York: Public Affairs.

Detroit Free Press. 1872. Going for Wolves. *Hawaiian Gazette* (27 March).

———. 1892. A Hunting Episode. *Atchison Globe* (29 Nov.).

Diamond, Joseph E., Thomas Amorosi, and David Perry. 2016. Late Woodland Subsistence at the Wolfersteig Site: A Multi-Component Site on the Esopus Creek. *Archaeology of Eastern North America* 44 (1): 131–160.

Dickerson, A.J. 1987. Rising Demand for Meat Spawns New Industry: Alligator Ranching. *Los Angeles Times*, 24 May.

Dickie, Gloria. 2018. When Cattle Go Missing in Wolf Territory, Who Should Pay the Price? *High Country News.* https://www.hcn.org/issues/50.12/wolves-when-cattle-go-missing-in-wolf-territory-who-should-pay-the-price.

Dickinson, Nate. 1993. *Common Sense Wildlife Management: Discourses on Personal Experiences.* Altamont, NY: Settle Hill Publishing.

Disagreeable Experience. 1892. *Portland Oregonian* (28 Nov.), 6.

Doerr, Michelle L., Jay B. Aninch, and Ernie P. Wiggers. 2001. Comparison of 4 Methods to Reduce White-Tailed Deer Abundance in an Urban Community. *Wildlife Society Bulletin* 29 (4): 1105–1113.

Dudley, Paul. 1720–1721. An Account of a Method Lately Found Out in New-England, for Discovering Where the Bees Hive in the Woods, in Order to Get Their Honey. *Philosophical Transactions* 31: 148–150.

Dudley, Paul, and John Chamberlayne. 1720–1721. A Description of the Moose-Deer in America. *Philosophical Transactions* 31: 165–168.

Dunaway, Wilma A. 1994. The Southern Fur Trade and the Incorporation of Southern Appalachia into the World-Economy, 1690–1763. *Fernand Braudel Center Review* 17 (2): 215–242.

Dunbar, Gary S. 1962. Deer-Keeping in Early South Carolina. *Agricultural History* 36 (2): 108–109.

Dunraven, Lord. 1879. Moose and Cariboo Hunting in Colorado and Canada. *Journal of the American Geographical Society of New York* 11: 334–368.

———. 1881. Hunting the Moose. *Daily Central City Register* (16 March).

Earle, Peter C. 2020. *Coronavirus and Human Rights.* Great Barrington: American Institute for Economic Research.

Eddy, J.W. 1924. *Hunting on Kenai Peninsula.* Seattle: Lowman & Hanford Co.

Edwards, George. 1753–1754. A Letter to Mr. Peter Collinson, F.R.S. Concerning the Pheasant of Pensylvania [sic], and the Otis Minor. *Philosophical Transactions* 48: 499–503.

Eisenhower, Dwight D. 1961. Farewell Address. https://www.ourdocuments.gov/doc.php?doc=90&page=transcript.

Erie Dispatch. 1884. Hunting in Pennsylvania. *Cleveland Daily Herald* (13 Jan.), 12.

Exchange. 1888. Sylvester Scott, Bear Hunter. *Milwaukee Journal* (16 March).

F.E.S. 1878. Hunting Experience. *Daily Evening Bulletin* (27 Feb.).

Feldpausch-Parker, Andrea, Israel D. Parker, and Elizabeth S. Vidon. 2017. Privileging Consumptive Use: A Critique of Ideology, Power, and Discourse in the North American Model of Wildlife Conservation. *Conservation and Society* 15 (1): 33–40.

Ferril, Will C. 1893. The Hunting Season. *Rocky Mountain News* (1 Oct.), 9.

Feyerabend, Paul. 2011. *The Tyranny of Science*. Cambridge: Polity Press.

Feynman, Richard. 1969. What Is Science? *The Physics Teacher* 7 (6): 313–320.

Fiedel, Stuart J. 2001. What Happened in the Early Woodland? *Archaeology of Eastern North America* 29 (1): 101–142.

Fischer, Hank. 2001. Who Pays for Wolves? *PERC* 19: 4.

Forest and Stream. 1882. The Trapper's Last Shot. *Daily Inter Ocean* (1 May), 10.

———. 1884. The Trapper's Last Shot. *Daily Inter Ocean* (1 May), 10.

———. 1889. The Bear-Hunting of Today. *Rocky Mountain News* (29 Nov.), 2.

———. 1892. Hunting with a Camera. *Daily Inter Ocean* (14 May), 2.

———. 1894. Corner in Elk Teeth. *Milwaukee Daily Sentinel* (28 Oct.), 13.

———. 1895. Trappers as Packers. *Milwaukee Daily Sentinel* (15 Dec.), 13.

———. 1896a. The Mastodon Not Extinct. *Milwaukee Journal* (27 Nov.).

———. 1896b. Trapping in Wisconsin. *Emporia Daily Gazette* (1 Feb.).

Formaini, Robert. 1990. *The Myth of Scientific Public Policy*. New Brunswick, NJ: Transaction Publishers.

Fort Collins Express. 1881. One Season's Hunting. *St. Louis Globe-Democrat* (2 Jan. 1881), 11.

Foster, H. Thomas, III, and Arthur D. Cohen. 2007. Palynological Evidence of the Effects of the Deerskin Trade on Forest Fires During the Eighteenth Century in Southeastern North America. *American Antiquity* 72 (1): 35–51.

Freese, Curtis H. 1997. The 'Use It or Lose It' Debate: Issues of a Conservation Paradox. In *Harvesting Wild Species: Implications for Biodiversity Conservation*, ed. Curtis H. Freese. Baltimore: Johns Hopkins University Press.

Fryxell, John M., David J.T. Hussell, Anne B. Lambert, and Peter C. Smith. 1991. Time Lags and Population Fluctuations in White-Tailed Deer. *Journal of Wildlife Management* 55 (3): 377–385.

Gainesville Florida Eagle. 1880. Trapping Beavers in North Florida. *Daily Evening Bulletin* (7 Apr.).

Gallardo, Julio C. 2018. Jurisdictions in Mexico. In *North American Wildlife Policy and Law*, ed. Bruce D. Leopold, Winifred B. Kessler, and James L. Cummins. Boone and Crockett Club: Missoula.

Gallman, Robert E., and Paul W. Rhode. 2019. *Capital in the Nineteenth Century*. Chicago: University of Chicago Press.

Garshells, David L., and Hank Hristienko. 2006. State and Provincial Estimates of American Black Bear Numbers Versus Assessments of Population Trend. *Ursus* 17 (1): 1–7.

Gillespie, Alexander. 2005. *Whaling Diplomacy: Defining Issues in International Law*. New York: Edward Elgar.

Glenwood Daily Avalanche. 1891. Slaughtering Game. *Rocky Mountain News* (23 Aug.), 12.

Golden, Katherine E., M. Nils Peterson, Christopher S. DePerno, Robert E. Bardon, and Christopher E. Moorman. 2013. Factors Shaping Private Landowner Engagement in Wildlife Management. *Wildlife Society Bulletin* 37 (1): 94–100.

Gooden, Jennifer, and Michael 't Sas-Rolfes. 2020. A Review of Critical Perspectives on Private Land Conservation in Academic Literature. *Ambio* 49 (1): 019–034.

Graham, John D., Laura C. Green, and Marc J. Roberts. 1988. *In Search of Safety: Chemical and Cancer Risk*. Cambridge: Harvard University Press.

Griffiths, Huw I., and David H. Thomas. 1997. *The Conservation and Management of the European Badger (Meles meles)*. Strasbourg: Council of Europe.

Grovenburg, Troy W., Christopher C. Swanson, Christopher N. Jacques, Christopher S. Deperno, Robert W. Klaver, Jonathan A. Jenks. 2011. Female White-Tailed Deer Survival Across Ecoregions in Minnesota and South Dakota. *American Midland Naturalist* 165 (2): 426–435.

Hairr, John. 2011. John Lawson's Observations on the Animals of Carolina. *North Carolina Historical Review* 88 (3): 312–332.

Harper's Magazine. 1870. Hunting the Canvas-Back. *Lowell Daily Citizen* (3 Feb.).

Harris, Larry D. 1984. *The Fragmented Forest: Island Biogeography Theory and the Preservation of Biotic Diversity*. Chicago: University of Chicago Press.

Hayek, Friedrich. 1945. The Use of Knowledge in Society. *American Economic Review* 35 (4): 519–530.

Helena Independent. 1896. Hunting Elk in Montana. *Galveston Daily News* (5 June).

Hennessey, O.T. 1895. Fawn Trappers in Trouble. *Daily Inter Ocean* (14 May).

Hodak, Marc, and Jack Masterson. 2021. *The Seeds of Their Own Destruction: Lessons from Utopian Experiments in Nineteenth-Century America, Part 1—The Other American Dream*. SSRN Working Paper.

Holcombe, Randall. 2019. *Liberty in Peril: Democracy and Power in American History*. San Francisco: Independent Institute.

Holmes, Thomas. 1893. Hunting the Polecat. *Scientific American* 68 (14): 218.

Holyoke, John. 2019. Maine Hunters on Track to Harvest More than 30,000 Deer This Year. *Bangor Daily News* (21 Nov.).

Honneland, Geir. 2013. *Making Fishery Agreements Work: Post-Agreement Bargaining in the Barents Sea*. New York: Edward Elgar.

Hough, E. 1889. Deer Coursing with Greyhounds. *Outing: An Illustrated Monthly Magazine of Sport, Travel and Recreation* 14: 426–435.

Hriestienko, Hank, and John E. McDonald. 2007. Going into the 21st Century: A Perspective on Trends and Controversies in the Management of the American Black Bear. *Ursus* 18 (1): 72–88.

Huffman, James L. 1995. In the Interests of Wildlife: Overcoming the Tradition of Public Rights. In *Wildlife in the Marketplace: The Political Economy Forum*, ed. Terry L. Anderson and Peter J. Hill. New York: Rowman & Littlefield.

Huggins, Laura E. 2013. *Environmental Entrepreneurship: Markets Meet the Environment in Unexpected Places*. Northampton, MA: Edward Elgar.

Hunt with the Yankton Sioux. 1873. Indian Hunting. *Lowell Daily Citizen* (22 Sept.).

Imperio, Simona, Massimiliano Ferrante, Alessandra Grignetti, Giacomo Santini, and Stefano Focardi. 2010. Investigating Population Dynamics in Ungulates: Do Hunting Statistics Make Up a Good Index of Population Abundance? *Wildlife Biology* 16: 205–214.

Indianapolis Journal. 1886. Trappers' Earnings Fifty Years Ago. *St. Louis Globe-Democrat* (4 Nov.), 10.

Ioannidis, John P.A. 2005. Why Most Published Research Findings Are False. *PLOS Medicine*. https://doi.org/10.1371/journal.pmed.0020124.

Isenberg, Andrew C. 2000. *The Destruction of the Bison, 1750–1920*. New York: Cambridge University Press.

Jahn, Laurence R. 1991. Foreword. In *Wildlife and Habitats in Managed Landscapes*, ed. Jon E. Rodiek and Eric G. Bolen. Washington, DC: Island Press.

Joanen, Ted, Larry McNease, Ruth M. Elsey, and Mark Staton. 1997. The Commercial Consumptive Use of the American Alligator (Alligator Mississippiensis) in Louisiana. In *Harvesting Wild Species: Implications for Biodiversity Conservation*, ed. Curtis H. Freese. Baltimore: Johns Hopkins University Press.

Johnsen, D. Bruce. 2009. Salmon, Science, and Reciprocity on the Northwest Coast. *Journal of Ecology and Society* 14 (43). http://www.ecologyandsociety.org/vol14/iss2/art43.

———. 2016. Reciprocity and Fractional Reserve Banking Among Northwest Coast Tribes. In *Unlocking the Wealth of Indian Nations*, ed. Terry Anderson. New York: Lexington.

Johnston, John, J.W. Ormsby, H.N. Campbell, Edward Silverman, H.T. Drake, Stephen Meunier, C.H. Mathews, Alfred James, and R.G. Richter. 1896. The Hunting of Deer. *Milwaukee Journal* (7 Nov.).

Jones, Hugh. 1699. Part of a Letter from the Reverend Mr. Hugh Jones to the Reverend Dr. Benjamin Woodroofe, F.R.S. Concerning Several Observables in Maryland. *Philosophical Transactions* 21: 436–442.

Jonker, Sandra A., Robert M. Muth, John F. Organ, Rodney R. Zwick, and William F. Siemer. 2006. Experiences with Beaver Damage and Attitudes of Massachusetts Residents Toward Beaver. *Wildlife Society Bulletin* 34 (4): 1009–1021.

Kansas City Star. 1890. Hunting Wild Ponies. *Bismarck Tribune* (19 Nov.), 2.

Kay, Charles E. 1997. Aboriginal Overkill and the Biogeography of Moose in Western North America. *Alces* 33: 141–164.

Kellogg, Vernon. 1926. Hunting Bighorn with a Camera. *Scientific Monthly* 23 (2): 112–116.

Kessler, Winifred B. 2018. The Canadian Constitution and Wildlife Policy. In *North American Wildlife Policy and Law*, ed. Bruce D. Leopold, Winifred B. Kessler, and James L. Cummins. Boone and Crockett Club: Missoula.

King, J.S. 1899. Lively Experience with an Old Hunter with Bears. *Denver Evening Post* (7 May), 16.

Kinietz, Vernon. 1940. Notes on the Algonquian Family Hunting Ground System. *American Anthropologist* 42 (1): 179.

Kluender, Richard A., Philip A. Tappe, and Michael E. Cartwright. 1992. Long-Term White-Tailed Deer Harvest Trends for the Southcentral United States. *Journal of the Arkansas Academy of Science* 46 (5): 49–52.

Knapp, Charles. 1935. Hogs Roman and Modern Boar Hunting, Ancient and Modern. *Classical Weekly* 28 (11): 81–84.

Knox, W. Matt. 2011. The Antler Religion. *Wildlife Society Bulletin* 35 (1): 45–48.

Koch, Paul L., and Anthony D. Barnosky. 2006. Late Quaternary Extinctions: State of the Debate. *Annual Review of Ecology, Evolution, and Systematics* 37: 215–250.

Kreppel, Peter. 1897. About Duck Hunting. *Milwaukee Journal* (17 Feb.), 8.

Kuhn, Thomas. 1996. *The Structure of Scientific Revolutions*. 3rd ed. Chicago: Chicago University Press.

Laliberte, Andrea S., and William J. Ripple. 2003. Wildlife Encounters by Lewis and Clark: A Spatial Analysis of Interactions Between Native Americans and Wildlife. *BioScience* 53 (10): 994–1003.

———. 2004. Range Contractions of North American Carnivores and Ungulates. *BioScience* 54 (2): 123–138.

Laskow, Sarah. 2017. The Giant Frog Farms of the 1930s Were a Giant Failure. *Atlas Obscura* (25 Oct.)

Leal, Donald R. 1998. Cooperating on the Commons: Case Studies in Community Fisheries. In *Who Owns the Environment?* ed. Peter J. Hill and Roger E. Meiners. New York: Rowman & Littlefield.

Lebel, Francois, Christian Dussault, Ariane Masse, and Steve D. Cote. 2012. Influence of Habitat Features and Hunter Behavior on White-Tailed Deer Harvest. *Journal of Wildlife Management* 76 (7): 1431–1440.

Leopold, Bruce D. 2018. History of Wildlife Policy and Law Through Colonial Times. In *North American Wildlife Policy and Law*, ed. Bruce D. Leopold, Winifred B. Kessler, and James L. Cummins. Boone and Crockett Club: Missoula.

Leopold, Bruce D., Winifred B. Kessler, and James L. Cummins. 2018. Preface. In *North American Wildlife Policy and Law*, ed. Bruce D. Leopold, Winifred B. Kessler, and James L. Cummins. Boone and Crockett Club: Missoula.

Lewiston Journal. 1881. Maine's Fur Industry. *St. Louis Globe-Democrat* (24 Dec.), 7.

———. 1885. A Bear Hunter. *Bismarck Tribune* (24 Oct.).

———. 1886. A Famous Maine Hunter. *St. Louis Globe-Democrat* (17 July), 4.

———. 1892. Trapping Bears and Killing Them. *Bismarck Tribune* (2 Apr.).

Liu, Nengye, Cassandra Brooks, and Tianbao Qin. 2019. *Governing Marine Living Resources in the Polar Regions.* New York: Edward Elgar.

Ljung, Per E., Shawn J. Riley, Thomas A. Heberlein, and Go Ran Ericsson. 2012. Eat Prey and Love: Game-Meat Consumption and Attitudes Toward Hunting. *Wildlife Society Bulletin* 36 (4): 669–675.

London Saturday Review. 1887. American Moose Hunting. *St. Louis Globe-Democrat* (14 Sept.), 6.

Lueck, Dean L. 1995. The Economic Organization of Wildlife Institutions. In *Wildlife in the Marketplace: The Political Economy Forum*, ed. Terry L. Anderson and Peter J. Hill. New York: Rowman & Littlefield.

Mack, Julie. 2017. Michigan Ranks No. 2 in 2016 Deer Harvest, and Other Deer-Hunting Facts. *Michigan Live* (6 Nov.).

MacLeod, William Christie. 1922. The Family Hunting Territory and Lenape Political Organization. *Anthropologist* 24 (4): 448–463.

Madrigal, T. Cregg, and Julie Zimmermann Holt. 2002. White-Tailed Deer Meat and Marrow Return Rates and Their Application to Eastern Woodlands Archaeology. *American Antiquity* 67 (4): 745–759.

Malley, Edward. 1895. Delightful Hunting Trip. *Boston Daily Advertiser* (17 Oct.), 7.

Malloy, Steven J. 2001. *Junk Science Judo: Self-Defense Against Health Scares & Scams.* Washington, DC: Cato Institute.

Mancall, Peter C. 2013. The Raw and the Cold: Five English Sailors in Sixteenth-Century Nanavut. *William and Mary Quarterly* 70 (1): 3–40.

Mancall, Peter C., and Thomas Weiss. 1999. Was Economic Growth Likely in Colonial British America? *Journal of Economic History* 59 (1): 17–40.

McCarty, George S. 1934. Scientific Methods of Game Breeding Will Make Good Hunting. *Scientific American* 151 (5): 234–235.

McHugh, J.L. 1977. Rise and Fall of World Whaling: The Tragedy of the Commons Illustrated. *Journal of International Affairs* 31 (1): 23–33.

Meiners, Roger, Pierre Desrochers, and Andrew Morriss, eds. 2012. *Silent Spring at 50: The False Crises of Rachel Carson.* Washington, DC: CATO.

Memphis Avalanche. 1883. Over Five Thousand Pigeons Killed. *St. Louis Globe-Democrat* (15 Dec.), 16.

Michelson, Truman. 1921. Note on the Hunting Territories of the Sauk and Fox. *American Anthropologist* 23 (2): 238–239.

Mighels, Philip V. 1897. Hints for Young Trappers. *Salt Lake Semi-Weekly Tribune* (11 May), 13.

Miles, Nelson A. 1895. Hunting Large Game. *North American Review* 161 (467): 484–492.

Millan, J. 1744. *The Present State of the Country and Inhabitants, Europeans and Indians, of Louisiana*. London: J. Millan.

Miller, Robert, Jr. 2012. *Reservation 'Capitalism': Economic Development in Indian Country*. New York: Praeger.

Moore, J.R. 1897. A Deer Hunting Law. *Milwaukee Daily Sentinel* (5 Feb.), 4.

Mountain Messenger. 1861. Trapping in Sierra County. *Daily Evening Bulletin* (15 Feb.).

Mulderink, Earl F. 2012. *New Bedford's Civil War*. New York: Fordham University Press.

Munkittrick, R.K. 1893. A Champion Hunter. *Yenowine's Illustrated News* (30 Dec.), 6.

Murdoch, W.G. Burn. 1917. *Modern Whaling and Bear Hunting*. London: Seeley, Service & Co.

Nagaoka, Lisa, Torben Rick, and Steve Wolverton. 2018. The Overkill Model and Its Impact on Research. *Ecology and Evolution* 8 (19): 9683–9696.

Natchez Free Trader. 1852. *Missouri Courier* (4 March).

Nesbit, William. 1926. *How to Hunt with the Camera*. London: G. Allen & Unwin Limited.

Neumann, Thomas W. 1985. Human-Wildlife Competition and the Passenger Pigeon: Population Growth from System Destabilization. *Human Ecology* 13 (4): 389–410.

New Orleans Time Democrat. 1883. Millions of Mallards. *St. Louis Globe-Democrat* (15 Dec.), 16.

New York Commercial. 1841. Venison. *North American* (27 Nov.).

———. 1875. Fox-Hunting. *St. Louis Globe-Democrat* (14 Nov.), 12.

New York Commercial Advertiser. 1860. The Buffalo Robe Trade. *Milwaukee Daily Sentinel* (24 Aug.).

New York Mail. 1887. Owl Hunting. *Boston Daily Advertiser* (25 Nov.), 5.

New York Mail and Express. 1886a. Hunting the Gray Squirrel. *St. Louis Globe-Democrat* (9 Dec.), 5.

———. 1886b. Points About Deer-Hunting. *St. Louis Globe-Democrat* (28 Dec.), 8.

New York Sun. 1878. Buffalo Hunting. *St. Louis Globe-Democrat* (24 Feb.), 12.

———. 1884. Hunting Notes. *St. Louis Globe-Democrat* (11 Aug.), 7.

———. 1888. Trapping Wild Beasts. *Milwaukee Journal* (11 Sept.).

————. 1889. Trapping Turkeys. *Atchison Daily Champion* (23 Jan.), 3.

————. 1897. New Hunting Rifles. *Milwaukee Journal* (5 Nov.), 7.

New York Times. 1877. Hunting in Florida. *St. Louis Globe-Democrat* (14 Jan.), 9.

————. 1884. The Hunters of Maine. *St. Louis Globe-Democrat* (10 May), 12.

————. 1885. An Apache Hunter. *Bismarck Tribune* (28 Apr.).

————. 1887. Trapping the Grizzly. *Daily Inter Ocean* (10 Nov.), 24.

New York Tribune. 1896. The Hunting Season. *Daily Inter Ocean* (22 Nov.), 27.

Nye, Bill. 1887. Fox Hunting. *Rocky Mountain News* (23 Oct.), 9.

O. 1889. Deer Hunting. *Portland Oregonian* (13 Jan.).

Organ, John F. 2018. The North American Model of Wildlife Conservation. In *North American Wildlife Policy and Law*, ed. Bruce D. Leopold, Winifred B. Kessler, and James L. Cummins. Boone and Crockett Club: Missoula.

Organ, John F., et al. 2012. The North American Model of Wildlife Conservation. *The Wildlife Society Technical Review* 12-04.

Organ, John F., Thomas A. Decker, and Tanya M. Lama. 2016. The North American Model and Captive Cervid Facilities—What Is the Threat? *Wildlife Society Bulletin* 40 (1): 10–13.

Ostrom, Elinor. 1990. *Governing the Commons: The Evolution of Institutions for Collective Action*. New York: Cambridge University Press.

————. 1992. The Rudiments of a Theory of the Origins, Survival, and Performance of Common-Property Institutions. In *Making the Commons Work: Theory, Practice, and Policy*, ed. Daniel W. Bromley. San Francisco: Institute for Contemporary Studies.

Ottaway, Andy. 2013. *Commercial Whaling. The Global Guide to Animal Protection*. Urbana: University of Illinois Press.

Oxley, J. Macdonald. 1888. Hunting the Moose. *Bismarck Tribune* (20 March).

Palmer, E. Laurence. 1939. Farm Forest Facts. *Cornell Rural School Leaflet* 33 (2): 1–32.

Patera, Pat. 1978. There's Big Money in the Secret Art of Frog Farming. *Mother Earth News*, July/Aug.

Pavao-Zuckerman, Barnet. 2007. Deerskins and Domesticates: Creek Subsistence and Economic Strategies in the Historic Period. *American Antiquity* 72 (1): 5–33.

Pellet, Frank. 1938. *History of American Beekeeping*. Ames: Iowa Collegiate Press.

Pennington, Mark. 2011. *Robust Political Economy: Classical Liberalism and the Future of Public Policy*. Northampton, MA: Edward Elgar.

Perttula, Timothy K., Cathy J. Crane, and James E. Bruseth. 1982. A Consideration of Caddoan Subsistence. *Southeastern Archaeology* 1 (2): 89–102.

Peterson, Markus J. 1992. Whalers, Cetologists, Environmentalists, and the International Management of Whaling. *International Organization* 46 (1): 147–186.

Peterson, Markus J., M. Nils Peterson, and Tarla Rai Peterson. 2016. What Makes Wildlife Wild?: How Identity May Shape the Public Trust Versus Wildlife Privatization Debate. *Wildlife Society Bulletin* 40 (3): 428–435.

Philadelphia Press. 1883. An Old Frog Hunter. *St. Louis Globe-Democrat* (12 Dec.), 6.

Pluckhahn, Thomas J., J. Matthew Compton, and Mary Theresa Bonhage-Freund. 2006. Evidence of Small-Scale Feasting from the Woodland Period Site of Kolomoki, Georgia. *Journal of Field Archaeology* 31 (3): 263–284.

Portland Oregonian. 1883. About Bears. *St. Louis Globe-Democrat* (10 Nov.), 4.

Pressly, Paul M. 2013. *On the Rim of the Caribbean: Colonial Georgia and the British Atlantic World.* Athens: University of Georgia Press.

Prukop, Joanna, and Ronald J. Regan. 2005. The Value of the North American Model of Wildlife Conservation: An International Association of Fish and Wildlife Agencies Position. *Wildlife Society Bulletin* 33 (1): 374–377.

Reagan, Albert B. 1919–1921. Hunting and Fishing of Various Tribes of Indians. *Transactions of the Kansas Academy of Science* 30: 443–448.

Reeves, Randall D., and Tim D. Smith. 2006. *Whales, Whaling, and Ocean Ecosystems.* Berkeley: University of California Press.

Reid, Colin T., and Walters Nsoh. 2016. *The Privatisation of Biodiversity? New Approaches to Conservation Law.* New York: Edward Elgar.

Reitz, Elizabeth J., and Gregory A. Waselkov. 2015. Vertebrate Use at Early Colonies on the Southeastern Coasts of Eastern North America. *International Journal of Historical Archaeology* 19 (1): 21–45.

Richards, John F. 2014. *The World Hunt: An Environmental History of the Commodification of Animals.* Berkeley: University of California Press.

Robinson, H.M. 1879. Indian Trappers, Hudson Bay. *Galveston Daily News* (28 May).

Rockwell, Robert H. 1922. Hunting the Big Brown Bear. *Brooklyn Museum Quarterly* 9 (1): 1–23.

———. 1923. Sheep Hunting in Alaska. *Brooklyn Museum Quarterly* 10 (2): 71–82.

———. 1924. Moose Hunting in Alaska. *Brooklyn Museum Quarterly* 11 (2): 76–81.

Rodiek, Jon E. 1991. Introduction. In *Wildlife and Habitats in Managed Landscapes*, ed. Jon E. Rodiek and Eric G. Bolen. Washington, DC: Island Press.

Roosevelt, Theodore. 1892. Antelope Hunting. *Daily Inter Ocean* (24 Jan.), 27.

Ross, W. Gillies. 1979. The Annual Catch of Greenland (Bowhead) Whales in Waters North of Canada, 1719–1915. *Arctic* 32 (2): 91–121.

Roth, Alvin. 2015. *Who Gets What And Why?: The Hidden World of Matchmaking and Market Design.* New York: William Collins.

Rucker, Randall, Walter N. Thurman, and Michael Burgett. 2019. Colony Collapse and the Consequences of Bee Disease: Market Adaptation to Environmental Change. *Journal of the Association of Environmental and Resource Economists* 6 (5): 927–960.

S. H. 1833. Bull-Hunting in Washitaw. *Maryland Gazette* (18 July).

San Francisco Call. 1885. Traps and Trappers. *Galveston Daily News* (5 Nov.), 6.

San Francisco Examiner. 1887. Trapping Black Bear. *St. Louis Globe-Democrat* (20 July), 6.

———. 1888. Hunting in Idaho. *Rocky Mountain News* (21 March), 3.

———. 1890. Hunting for Mud Turtles. *Bismarck Tribune* (9 Apr.).

Savage, James. 1825. *The History of New England from 1630 to 1649 by John Winthrop*. Boston: Phelps and Farnham.

Savage, Henry L. 1933. Hunting in the Middle Ages. *Speculum* 8 (1): 30–41.

Scanland, J.M. 1893. Hunting the Elk. *Bismarck Tribune* (23 July), 4.

Schorr, Robert A., Paul M. Lukacs, and Justin A. Gude. 2014. The Montana Deer and Elk Hunting Population: The Importance of Cohort Group, License Price, and Population Demographics on Hunter Retention, Recruitment, and Population Change. *Journal of Wildlife Management* 78 (5): 944–952.

Schuck, Peter. 2015. *Why Government Fails So Often: And How It Can Do Better*. Princeton: Princeton University Press.

Schwabe, Kurt A., and Peter W. Schuhmann. 2002. Deer-Vehicle Collisions and Deer Value: An Analysis of Competing Literatures. *Wildlife Society Bulletin* 30 (2): 609–615.

Seay, Katharyn. 2019. Alligator Mississippiensis. *Animal Diversity Web*. https://animaldiversity.org/accounts/Alligator_mississippiensis/.

Shaw, Christopher W. 2019. *Money, Power, and the People: The American Struggle to Make Banking Democratic*. Chicago: University of Chicago Press.

Shields, G. O. 1887. Hunting the Grizzly. *St. Louis Daily Globe-Democrat* (31 Jul.), 28.

Shoemaker, Nancy. 2005. Whale Meat in American History. *Environmental History* 10 (2): 269–294.

Shrader-Frechette, K.S. 1991. *Risk and Rationality: Philosophical Foundations for Populist Reforms*. Berkeley: University of California Press.

Simon, Julian. 1999. *Hoodwinking the Nation*. New Brunswick: Transaction Publishers.

Smalley, Andrea L. 2016. 'They Steal Our Deer and Land': Contested Hunting Grounds in the Trans-Appalachian West. *Register of the Kentucky Historical Society* 114 (3/4): 303–339.

Smith, De Cost. 1889. Onondaga Superstitions. Hunting. *Journal of American Folklore* 2 (7): 282–283.

Smith, Christian A. 2011. The Role of State Wildlife Professionals Under the Public Trust Doctrine. *Journal of Wildlife Management* 75 (7): 1539–1543.

Smith, Paul A. 2019. Deer Population at Potential Record High as Hunters Set Sights on 2019 Wisconsin Gun Season. *Milwaukee Sentinel Journal* (16 Nov.).

Smith, Andrew, and Robert E. Wright. 2021. Sowing the Seeds of a Future Crisis: The SEC and the Emergence of the Nationally Recognized Statistical Rating Organization (NRSRO) Category, 1971–75. *Business History Review* 95 (4): 739–64. https://doi.org/10.1017/S0007680521000106.

Sokos, Christos K., M. Nils Peterson, Periklis K. Birtsas, and Nikolas D. Hasanagas. 2014. Insights for Contemporary Hunting from Ancient Hellenic Culture. *Wildlife Society Bulletin* 38 (3): 451–457.

Southwell, Robert. 1686–1692. The Method the Indians in Virginia and Carolina Use to Dress Buck and Doe Skins. *Philosophical Transactions* 16: 532–533.

Sowell, Thomas. 2009. *Intellectuals and Society*. New York: Perseus Group.

Special Correspondence. 1887. Deer Hunting. *St. Louis Globe-Democrat* (7 Aug.), 20.

Speck, Frank G. 1915. The Family Hunting Band as the Basis of Algonkian Social Organization. *American Anthropologist* 17 (2): 289–305.

———. 1923. Mistassini Hunting Territories in the Labrador Peninsula. *American Anthropologist* 25 (4): 452–471.

Speck, Frank G., and Loren C. Eiseley. 1939. Significance of Hunting Territory Systems of the Algonkian in Social Theory. *American Anthropologist* 41 (2): 269–280.

———. 1942. Montagnais-Naskapi Bands and Family Hunting Districts of the Central and Southeastern Labrador Peninsula. *Proceedings of the American Philosophical Society* 85 (2): 215–242.

Spiess, Arthur, Kristin Sobolik, Diana Crader, John Mosher, and Deborah Wilson. 2006. Cod, Clams and Deer: The Food Remains from Indiantown Island. *Archaeology of Eastern North America* 34 (1): 1471–1487.

Sporting Journal. 1885. Hunting Wild Turkeys. *Los Angeles Times* (31 May).

St. Louis Globe-Democrat. 1894. Hunting Wild with Tame Turkeys. *Bismarck Tribune* (26 Feb.), 2.

———. 1895. Hunting Dakota Wolves. *Daily Inter Ocean* (2 June), 31.

St. Paul Pioneer Press. 1893. The Successful Hunter. *Bangor Daily Whig and Courier* (8 Dec.).

Staunton Vindicator. 1877. A Great Hunter Killed. *St. Louis Globe-Democrat* (17 Dec.), 2.

Stothers, David M., and Timothy J. Abel. 1993. Archaeological Reflections of the Late Archaic and Early Woodland Time Periods in the Western Lake Erie Region. *Archaeology of Eastern North America* 21 (1): 25–109.

Swanson, Drew A. 2018. *Beyond the Mountains: Commodifying Appalachian Environments*. Athens: University of Georgia Press.

Taylor, H.W. 1888. Turkey Hunting. *Milwaukee Daily Sentinel* (1 Jan.).

Thomas, William S. 1906. *Hunting Big Game with Gun and with Kodak: A Record of Personal Experiences in the United States, Canada, and Mexico.* New York: G.P. Putnam's Sons.

Tobias, Terry N., and James J. Kay. 1994. The Bush Harvest in Pinehouse, Saskatchewan, Canada. *Arctic* 47 (3): 207–221.

Toledo Rep. 1854. Venison. *Bangor Daily Whig and Courier* (25 Jan.).

Trapper's Guide. 1865. *Bangor Daily Whig and Courier* (30 Sept.).

Trefethen, James B. 1975. *An American Crusade for Wildlife.* New York: Winchester Press.

Troyer, Dianna. 2019. Western Innovator: Frog 'Ranch' Keeps Owners Hopping. *Capital Press*, 12 July.

Usner, Daniel H., Jr. 1985. The Deerskin Trade in French Louisiana. *Proceedings of the Meeting of the French Colonial Historical Society* 10: 75–93.

———. 1992. *Indians, Settlers, and Slaves in a Frontier Exchange Economy: The Lower Mississippi Valley Before 1783.* Chapel Hill: University of North Carolina Press.

Vantassel, Stephen M., Tim L. Hiller, Kelly D.J. Powell, and Scott E. Hyngstrom. 2010. Using Advancements in Cable-Trapping to Overcome Barriers to Furbearer Management in the United States. *Journal of Wildlife Management* 74 (5): 934–939.

Vercauteren, Kurt C., Charles W. Anderson, Timothy R. Van Deelen, W. David Drake, David Walter, Stephen M. Vantassel, and Scott E. Hyngstrom. 2011. Regulated Commercial Harvest to Manage Overabundant White-Tailed Deer: An Idea to Consider? *Wildlife Society Bulletin* 35 (3): 185–194.

Vicksburg Commercial Herald. 1892. Hunting the Hunter. *New Orleans Daily Picayune* (7 Nov. 1892), 7.

Vrtiska, Mark P., James H. Gammonley, Luke W. Naylor, and Andrew H. Raedeke. 2013. Economic and Conservation Ramifications from the Decline of Waterfowl Hunters. *Wildlife Society Bulletin* 37 (2): 380–388.

W. S. 1915. McIlhenny's The Wild Turkey and Its Hunting. *The Auk* 32 (1): 115–116.

———. 1924. *Recollections of Fifty Years Hunting and Shooting* by Wm. B. Mershon. *The Auk* 41 (2): 365.

Wagner, Richard E. 1998. The Constitutional Protection of Private Property. In *Who Owns the Environment?* ed. Peter J. Hill and Roger E. Meiners. New York: Rowman & Littlefield.

Wales, William. 1770. Journal of a Voyage, Made by Order of the Royal Society, to Churchill River, on the North-west Coast of Hudson's Bay. *Philosophical Transactions* 60: 100–136.

Walsh, Virginia M. 1999. Illegal Whaling for Humpbacks by the Soviet Union in the Antarctic, 1947–1972. *Journal of Environment & Development* 8 (3): 307–327.

Warnock, Mary. 2015. *Critical Reflections on Ownership*. New York: Edward Elgar.

Waselkov, Gregory A. 1978. Evolution of Deer Hunting in the Eastern Woodlands. *Midcontinental Journal of Archaeology* 3 (1): 15–34.

Webb, G. Kent. 2018. Searching the Internet to Estimate Deer Population Trends in the U.S., California, and Connecticut. *Issues in Information Systems* 19 (2): 163–173.

West Bloomfield. 1889. An Adirondack Hunter. *Daily Inter Ocean* (22 Dec.)

Western Hunter. 1889. Hunting Buffaloes. *Milwaukee Sentinel* (30 Aug.), 4.

Weyawega Times. 1872. An Old Trapper's Adventure. *Lowell Daily Citizen* (2 Nov.).

Wheat, Joe Ben, Harold E. Malde, and Estella B. Leopold. 1972. The Olsen-Chubbuck Site: A Paleo-Indian Bison Kill. *Memoirs of the Society for American Archaeology* 26: 1–180.

Wheeler, E.P. 1930. Journeys About Nain Winter Hunting with the Labrador Eskimo. *Geographical Review* 20 (3): 454–468.

Whig Man. 1899. Hunting Season. *Bangor Daily Whig and Courier* (9 May), 2.

White, Richard. 1983. *The Roots of Dependency: Subsistence, Environment, and Social Change Among the Choctaws, Pawnees, and Navajos*. Lincoln: University of Nebraska Press.

Whitney, Leon F. 1931. The Raccoon and Its Hunting. *Journal of Mammalogy* 12 (1): 29–38.

Williams, M.C. 1889. Happy Hunting Grounds. *Daily Inter Ocean* (1 Sept.), 21.

Williams, A., and D. Bugbee. 1865. The Trapper's Guide. *Bangor Daily Whig and Courier* (19 Sept.).

Winkler, Richelle, and Keith Warnke. 2013. The Future of Hunting: An Age-Period-Cohort Analysis of Deer Hunter Decline. *Population and Environment* 34 (4): 460–480.

Winnebago County Press. 1870. The Muskrat Catch. *Milwaukee Daily Sentinel* (10 May).

Winona Republican. 1873. The Pigeon Business. *Milwaukee Daily Sentinel* (23 May), 2.

Wright, Charles. 1868a. Bears and Bear-Hunting. *The American Naturalist* 2 (3): 121–124.

———. 1868b. Deer and Deer-Hunting in Texas. *The American Naturalist* 2 (9): 466–476.

Wright, Robert E. 2010. *Bailouts: Public Money, Private Profit*. New York: Columbia University Press.

———. 2019. America's Fur Business, Parts I, II, and III. *Fur Traders & Rendezvous*. https://www.alfredjacobmiller.com/explore/americasfur business1/.

———. 2022. The Political Economy of Modern Wildlife Management: How Commercialization Could Reduce Game Overabundance. *Independent Review* 26 (4): 512–32.

Wright, Robert E., and Thomas Zeiler, eds. 2014. *Guide to U.S. Economic Policy.* Washington: CQ Press.

Young, Kimball, and Thomas D. Cutsforth. 1928. Hunting Superstitions in the Cow Creek Region of Southern Oregon. *Journal of American Folklore* 41 (160): 283–285.

Zink, Robert M. 2014. *The Three-Minute Outdoorsman: Wild Science from Magnetic Deer to Mumbling Carp.* St. Paul: University of Minnesota Press.

INDEX[1]

[1] Note: Page numbers followed by 'n' refer to notes.

© The Author(s), under exclusive license to Springer Nature
Switzerland AG 2022
R. E. Wright, *The History and Evolution of the North American
Wildlife Conservation Model*,
https://doi.org/10.1007/978-3-031-06163-9